Rethinking Victorian Culture

Also edited by Juliet John and Alice Jenkins

* REREADING VICTORIAN FICTION

Also edited by Juliet John

CULT CRIMINALS
The Newgate Novels

* *From the same publishers*

Rethinking Victorian Culture

Edited by

Juliet John
University of Salford

and

Alice Jenkins
University of Glasgow

Foreword by John Sutherland

First published in Great Britain 2000 by
MACMILLAN PRESS LTD
Houndmills, Basingstoke, Hampshire RG21 6XS and London
Companies and representatives throughout the world

A catalogue record for this book is available from the British Library.

ISBN 0–333–71446–6

 First published in the United States of America 2000 by
ST. MARTIN'S PRESS, INC.,
Scholarly and Reference Division,
175 Fifth Avenue, New York, N.Y. 10010

ISBN 0–312–22679–9

Library of Congress Cataloging-in-Publication Data
Rethinking Victorian culture / edited by Juliet John and Alice Jenkins ; foreword by John Sutherland.
p. cm.
Includes bibliographical references and index.
ISBN 0–312–22679–9
1. English literature—19th century—History and criticism.
2. Great Britain—History—Victoria, 1837–1901. 3. Great Britain–
–Civilization—19th century. I. John, Juliet, 1967– .
II. Jenkins, Alice, 1970– .
PR461.R48 1999
820.9'008—dc21   99–29985
                                                                   CIP

Selection, editorial matter and Chapter 1 © Juliet John and Alice Jenkins 2000
Foreword © John Sutherland 2000
Chapter 2 © Stefan Collini 2000
Chapter 5 © Stephen Prickett 2000
Chapter 6 © Andrew Sanders 2000
Chapter 9 © Juliet John 2000
Chapter 11 © Alice Jenkins 2000
Chapters 3, 4, 7, 8, 10, 12–15 © Macmillan Press Ltd 2000

All rights reserved. No reproduction, copy or transmission of this publication may be made without written permission.

No paragraph of this publication may be reproduced, copied or transmitted save with written permission or in accordance with the provisions of the Copyright, Designs and Patents Act 1988, or under the terms of any licence permitting limited copying issued by the Copyright Licensing Agency, 90 Tottenham Court Road, London W1P 0LP.

Any person who does any unauthorised act in relation to this publication may be liable to criminal prosecution and civil claims for damages.

The authors have asserted their rights to be identified as the authors of this work in accordance with the Copyright, Designs and Patents Act 1988.

This book is printed on paper suitable for recycling and made from fully managed and sustained forest sources.

10 9 8 7 6 5 4 3 2 1
09 08 07 06 05 04 03 02 01 00

Printed and bound in Great Britain by Antony Rowe Ltd, Chippenham, Wiltshire

For Calum Forsyth and Stephen Swann

KING ALFRED'S COLLEGE
WINCHESTER

820.9008
/JOH   KA02007371

Contents

List of Illustrations	ix
Acknowledgements	x
Notes on the Contributors	xi
Foreword by John Sutherland	xv

1. Introduction   1
   Juliet John and Alice Jenkins

2. From 'Non-Fiction Prose' to 'Cultural Criticism': Genre and Disciplinarity in Victorian Studies   13
   Stefan Collini

3. Republican versus Victorian: Radical Writing in the Later Years of the Nineteenth Century   29
   John Lucas

4. 'The Mote Within the Eye': Dust and Victorian Vision   46
   Kate Flint

5. Purging Christianity of its Semitic Origins: Kingsley, Arnold and the Bible   64
   Stephen Prickett

6. Dickens and the Millennium   81
   Andrew Sanders

7. Re-Arranging the Year: the Almanac, the Day Book and the Year Book as Popular Literary Forms, 1789–1860   92
   Brian Maidment

8. Tennyson and the Apostles   115
   John Coyle and Richard Cronin

9. *Twist*ing the Newgate Tale: Dickens, Popular Culture and the Politics of Genre   127
   Juliet John

10	'More interesting than all the books, save one': Charles Kingsley's Construction of Natural History *Francis O'Gorman*	147
11	Writing the Self and Writing Science: Mary Somerville as Autobiographer *Alice Jenkins*	163
12	A Taste for Change in *Our Mutual Friend*: Cultivation or Education? *Pam Morris*	180
13	'The Mother of our Mothers': Ghostly Strategies in Women's Writing *Anthea Trodd*	196
14	Actresses, Autobiography and the 1890s *Gail Marshall*	209
15	Modernity and Progress in Economics and Aesthetics *Regenia Gagnier*	223
Index		240

List of Illustrations

1 Extra illustration to Dickens's *Pickwick Papers* (Chapman and Hall 1837), possibly by Thomas Onwhyn. Onwhyn's pirated illustrations were published by E. Grattan.   97

2 *Social Reformers' Almanac* for 1841. Illustration by courtesy of the Museum of Labour History, Manchester.   99

3 The title-page of *The British Almanac* for 1848. Published by Charles Knight for the Society for the Diffusion of Useful Knowledge, *The British Almanac* was deliberately aimed at replacing predictive and superstitious almanacs with rational information for artisan readers.   102

4 Title-page and opening spread from *The Instructor*, a five-volume publication aimed at artisan readers and published by J. W. Parker on behalf of the Society for Promoting Christian Knowledge. First published in the 1830s, *The Instructor* was extensively reprinted in the 1840s and 1850s. Volume IV contains extensive discussion of the calendar and the origin and uses of the almanac.   104

5 Title-page spread from William Hone's *The Everyday Book*, reprinted by Thomas Tegg in 1830.   105

6 Title-page from George Cruikshank's *The Comic Almanac for 1840*, published by Charles Tilt.   107

7 A page from the *Punch Almanac* for 1842.   109

8 The opening page of *Chambers's Book of Days*, published by W. and R. Chambers (Edinburgh, n.d. [1862]).   110

Acknowledgements

We would first like to thank the Department of English at the University of Liverpool for supporting the conference, 'Victorian Studies: into the Twenty-First Century' (September 1996), out of which this anthology grew. In particular, we wish to thank Blackwell's bookshop and the following individuals for their practical help and support with the organization of the conference: Miriam Allott, Jagori Bandyopadhyay, Mary Clinton, Lucy Crispin, Philip Davis, Kelvin Everest, Val Gough, Liz Hedgecock, Jo Knowles, David Mills, Cathy Rees, Susan Rowland, Jill Rudd, Karl Simms and Barbara Smith.

Since leaving the University of Liverpool, we have both received assistance during the publication process from friends and colleagues at our new institutions and elsewhere. We owe special thanks to Helen Hackett (UCL) and Dominic Rainsford (University of Aarhus) for advice in the early stages, to Hazel Clarke (Edge Hill University College, Lancashire) for administrative support, and to Desmond O'Brien (University of Glasgow) for the generosity, time and patience he devoted to solving our computer problems. At the University of Salford, Antony Rowland and Avril Horner have both been attentive and patient readers of sections of this book, and Peter Buse has answered particular queries uncomplainingly.

Finally, we are grateful to Charmian Hearne for her unfailing patience, support and understanding throughout this project.

Note: The editors and publishers would like to thank Cambridge University Press for permission to reproduce extracts from *Actresses on the Victorian Stage* by Gail Marshall (1998) in Chapter 14.

Notes on the Contributors

Stefan Collini is Reader in Intellectual History and English Literature at Cambridge University. Among his publications are *Public Moralists: Political Thought and Intellectual Life in Britain, 1850–1930* (1991), *Matthew Arnold: a Critical Portrait* (1994), and *English Pasts* (1999). He is currently working on a study of intellectuals in twentieth-century Britain.

John Coyle and Richard Cronin teach English at the University of Glasgow. John Coyle works on Ruskin, Proust and Modernism, and Richard Cronin works on Romantic and Victorian literature.

Kate Flint is Reader in English and Fellow of Linacre College, Oxford University. She has published *The Woman Reader, 1837–1914* (1993), and many articles on Victorian and twentieth-century literature, painting and cultural history. *The Victorians and the Visual Imagination* is forthcoming, and her current research focuses on the Victorians and Native Americans.

Regenia Gagnier is Professor of English at the University of Exeter, where she teaches Victorian Studies, social theory, feminist theory and interdisciplinary studies. Her books include *Idylls of the Marketplace: Oscar Wilde and the Victorian Public* (1986), *Subjectivities: a History of Self-Representation in Britain, 1832–1920* (1991), and an edited collection, *Critical Essays on Oscar Wilde* (1992). Most recently she has published essays on the histories of economics and aesthetics in market society in *Feminist Economics, Journal of Economic Issues, Victorian Studies* and *Political Theory*.

Alice Jenkins is a Lecturer in English Literature at the University of Glasgow. Her research focuses on Romantic and Victorian literature, particularly in relation to the history of science. She has published on Humphry Davy and Michael Faraday and is completing a study of spatial discourse in early-nineteenth-century literary and scientific writing. She is the co-editor (with Juliet John) of *Rereading Victorian Fiction* (1999).

Juliet John is a Lecturer in English at the University of Salford. She is the editor of *Cult Criminals: the Newgate Novels*, 6 vols (1998) and the co-editor (with Alice Jenkins) of *Rethinking Victorian Fiction* (1999). She has published articles mainly on Dickens and is currently completing a book on Dickens's villains and popular melodrama.

John Lucas is currently Research Professor at Nottingham Trent University, having previously spent 19 years as Professor of English at Loughborough University. Before going to Loughborough, he was Reader in English at the University of Nottingham. He has also been Lord Byron Visiting Professor of English at the University of Athens (1984–85) and a Visiting Fellow at La Trobe University, Melbourne (Feb.–April, 1992). Among his publications are *Tradition and Tolerance in 19th-Century Fiction* (with John Goode and David Howard, 1966), *The Melancholy Man: a Study of Dickens' Fiction* (1970; 2nd edn 1974), *The Literature of Change: Studies in the Provincial Novel* (1977; paperback 1978), *Romantic to Modern: Essays in Literature and Ideas* (1982), *England and Englishness: Ideas of Nationhood in English Poetry, 1688–1900* (1990; paperback 1991), *Dickens: the Major Novels* (1992), *John Clare: Writer and Work* (1994), *Writing and Radicalism* (ed., 1996), *The Radical 'Twenties: Politics, Writing, Culture* (1997), and (in the Critical Reader Series) *William Blake* (ed., 1998). He is also the author of three collections of poetry including *One for the Piano: Poems* (1997), and of versions of the poems of *Egil's Saga*, first published in1975, reprinted 1985 and 1993.

Brian Maidment is Professor and Head of English at the University of Huddersfield. He has wide interests in nineteeenth-century mass-circulation literature, periodicals, radical writing and popular culture. He has published *The Poorhouse Fugitives: Self-Taught Poets and Poetry in Victorian Britain* (1987) and *Reading Popular Prints* (1996). He is currently completing a wide-ranging study of the representation of dustmen in nineteenth-century plays, fiction and graphic satire.

Gail Marshall is a Lecturer in Victorian literature in the School of English at the University of Leeds. She has published articles on George Eliot, Ibsen, and late-Victorian theatre, and is the author *of Actresses on the Victorian Stage: Feminine Performance and the Galatea Myth* (1998). She is currently working on a study of Shakespeare and Victorian women.

Pam Morris is Professor of Modern Critical Studies at Liverpool John Moores University. She has written *Dickens's Class Consciousness: a Marginal View* (1991), *Literature and Feminism* (1993), and *The Bakhtin Reader* (1994), and has recently edited Elizabeth Gaskell's *Wives and Daughters*.

Francis O'Gorman read English as Organ Scholar of Lady Margaret Hall, Oxford, and wrote a DPhil on Ruskin there. He was Lecturer in English at Pembroke College, Oxford, 1993–96, and is now Research Fellow in English of Cheltenham and Gloucester College of Higher Education. He has published various articles on Ruskin in UK and US journals, and his book *Late Ruskin* is forthcoming. He has also published articles on Tyndall and Browning, and is co-editing a collection of essays on Margaret Oliphant.

Stephen Prickett is Regius Professor of English Language and Literature at the University of Glasgow. Previous appointments include the Chair of English at the Australian National University in Canberra (1983–89), and teaching posts at the Universities of Sussex (1967–82), Minnesota (1979–80), and Smith College, Massachusetts (1970–71). He is Fellow of the Australian Academy of the Humanities, former Chairman of the UK Higher Education Foundation, President of the European Society for the Study of Literature and Theology, and Patron of the George MacDonald Society. He has published more than fifty articles and ten books on Romanticism, Victorian studies and related topics, especially relating to literature and theology, including *Wordsworth: the Poetry of Growth* (1970), *Romanticism and Religion: the Tradition of Wordsworth and Coleridge in the Victorian Church* (1976), *Victorian Fantasy* (1978), *The Romantics* (ed., 1981), *Words and the Word: Language and Literary Theory* (ed., 1991) (with Robert Barnes), *The Bible* in the Landmarks of World Literature Series, *Origins of Narrative: the Romantic Appropriation of the Bible* (1996), and a World's Classics *Bible* (ed. with Robert Carroll, 1997).

Andrew Sanders is Professor of English in the University of Durham. He is the author of *The Victorian Historical Novel, Charles Dickens: Resurrectionist, The Companion to 'A Tale of Two Cities', The Short Oxford History of English Literature, Anthony Trollope* (in the Writers and their Work series) and *Dickens and the Spirit of the Age*. He was editor of the *Dickensian*, 1978–87, and he has edited novels by Dickens, Eliot, Gaskell, Thackeray and Hughes.

Anthea Trodd is Senior Lecturer in the Department of English, Keele University. She has written *Domestic Crime in the Victorian Novel* (1989), and *Women's Writing in English: Britain 1900–1945* (1998), edited *The Woman in White* and *The Moonstone*, and written articles on sensation fiction. She is currently working, with John Bowen and Deborah Wynne, on a project funded by the Leverhulme Trust on the collaborations of Dickens and Collins.

Foreword

When he heard the word 'culture', Goering is supposed to have said, he reached for his revolver. Victorianists of my generation may feel much the same when they hear the term 'cultural criticism'. Both *Rethinking Victorian Culture* and its partner volume, *Rereading Victorian Fiction*, were spun off from 'Victorian Studies: into the Twenty-first Century', a conference held at the University of Liverpool in 1996. The essays in this book are those that do not fit as easily into any single generic glass slipper as those in the partner volume do (although even there a few toes had to be mangled). They comprise 'history of ideas', 'new historicist criticism', 'non fiction prose studies', 'feminist analysis', old-fashioned 'contextualization' and – yes – cultural criticism. The general range of this collection is doctrinally wide and eclectic.

The first essay in *Rethinking Victorian Culture* is a blastingly powerful piece by Stefan Collini in which – as he has done elsewhere in the late 1990s – he takes on the whole cultural studies industry, specifically its recent transatlantic mutations. Collini's essay, which strikes me as a masterful overview (and an entertaining read to boot), traces origins, the dubious subsoil of the English department, and asserts his own sense of Cambridge pedigree. He ends with a rousing call to reorganize our priorities: 'we must adapt our generic categories to theirs [the writers of nineteenth-century non-fictional prose] rather than the other way round'. This is one of those essays which either changes its reader, or leaves her/him fuming.

'Culture' is a traditionally vexed and problematic term. Some of us feel safer with the other term in the title of this volume. John Lucas, however, is as stringently revisionist as Collini: 'There is a strong case', Lucas declares, 'for arguing that, except in the most rigorously controlled of contexts, "Victorian" and "Victorianism" are terms we could well do without.' They are employed, Lucas sternly maintains, 'all too frequently in ways that are chronologically indefensible, historically dubious, intellectually confusing, and ideologically unacceptable – at least if you're a socialist'. His essay on 'Radical Writing' is pointedly located in 'the Later Years of the Nineteenth Century' and the 'V-word' is studiously avoided in the subsequent exposition. Collini's and Lucas's papers would have made a superb basis for a two-day workshop

('Victorian Studies: The Way Forward!'). These two essays will, I suspect, reverberate for some time.

Kate Flint's cogitation on 'dust' was a dazzling dais performance, marked by lateral referential jumps which suggested hypertext as well as hyper-sophistication. Fiendishly clever as it was, the lecture made one think and repeatedly gave one the sense (momentarily at least) of thinking and perceiving like a Victorian. Other pieces in this volume find little jig-saw complementarity with each other but all are in their ways illuminating. They also witness to how aggressively, under the 'cultural studies' banner, 'English' has moved into borderlands with other subjects and disciplines. Stephen Prickett's meditation on race and Victorian biblical criticism can be seen as an incursion into historical anthropology as can Pam Morris's delightfully introduced essay about Max Müller's linguistics and *Our Mutual Friend*. Regenia Gagnier's essay on late-century modernity is informed by a specialist's expertise in Victorian economic theory and its current literary equivalent. One notes in this volume – in the essays of Alice Jenkins, Anthea Trodd and Gail Marshall – an organized awareness of women's literary and cultural activities.

I suspect that readers will pluck different good things out of this bran tub. If forced to nominate my good things I would have to say that I was most entertained in the lecture room by Kate Flint, Stephen Prickett and Andrew Sanders, and instructed on the page by all the contributors here represented. Overall, the achievement of the book, like that of its originating conference, is that it does not impose artificial or forced divisions.

JOHN SUTHERLAND

1
Introduction

Juliet John and Alice Jenkins

> If culture, then, is a study of perfection, and of harmonious perfection, general perfection, and perfection which consists in becoming something rather than in having something, in an inward condition of the mind and spirit, not in an outward set of circumstances, – it is clear that culture [...] has a very important function to fulfil for mankind.
>
> Matthew Arnold, *Culture and Anarchy* (1869)

> The [...] word, *culture*, similarly changes, in the same critical period [*c.* 1780–1850]. Before this period, it had meant, primarily, the 'tending of natural growth', and then, by analogy, a process of human training. But this latter use, which had usually been a culture *of* something, was changed, in the nineteenth century, to *culture* as such, a thing in itself. It came to mean, first, 'a general state or habit of the mind', having close relations with the idea of human perfection. Second, it came to mean 'the general state of intellectual development, in a society as a whole'. Third, it came to mean 'the general body of the arts'. Fourth, later in the century, it came to mean 'a whole way of life, material, intellectual and spiritual'. It came also, as we know, to be a word which often provoked either hostility or embarrassment.
>
> Raymond Williams, *Culture and Society, 1870–1950* (1958)

Culture and Anarchy and *Culture and Society* are arguably the two most important attempts to theorize culture by British intellectuals since the accession of Queen Victoria. The differences between the two critiques are, of course, infamous: Arnold, an élitist idealist, argues for the importance of culture in the moral and intellectual elevation of the individual and society; Williams, a materialist Marxist, argues for the importance of a 'common culture' in a fair and equal society.[1] Arnold sees culture as a defence against anarchy; Williams sees a 'common culture' as the possible safeguard of democracy. In *Culture and Anarchy*, the (cultured, intellectual) individual is an agent of social change; in *Culture and Society*, society – which has become divorced from 'culture' – effects individual change. There is agreement, however, about one thing: in the Victorian period, the meaning and function of 'culture' changes.

The similarities, perhaps surprisingly, do not end there. Both Arnold and Williams are equally troubled, from their very different wings of the political spectrum, by the separation of 'high culture' from 'culture as a whole way of life'. Arnold is motivated by concern about the peripheral role of the intellectual and high culture in the life of the nation (which is itself impoverished without the 'sweetness and light' that is culture);[2] Williams is troubled by the exclusion of the masses from the cultural capital that high culture represents. However, both recognize the Victorian period as witnessing new and distinctive contests about the relationship between 'high' and 'popular' culture, and at the same time experiencing a blurring of the boundaries defining ownership of these categories.

The radical changes in the social, political and economic structures of nineteenth-century Britain make this dislocation unsurprising if not predictable. Industrialization, the Reform Bills, the increase in literacy, and changes to the legal and economic structure of the nation are just some of the momentous historical developments which began or gathered pace in the early nineteenth century; these changes, and the shift in attitudes which attended them, have led some to regard this period as the beginning of a 'modern' era.[3] Quieter perhaps than the social, political and economic revolutions of the nineteenth century, but no less formative in the creation of a 'modern' state, was the cultural revolution which enabled more and more people to gain access to 'culture'. Developments in the publishing trade in the 1820s and 1830s meant that books and newspapers were expanding their readership, moving further down the social ladder in the early Victorian period than ever before. Until 1830, the cheap fiction market had largely been left to small and disreputable publishers such as the Minerva Press, but in June 1829 when Tom Cadell issued the Author's Edition of the Waverley novels in five-shilling volumes, 'he inaugurated the vogue of inexpensive recent fiction imprints'.[4] In 1831, Colburn and Bentley's

Standard Novels were published at six shillings each. These developments began a sharp fall in book prices, leading to a broader readership, a trend which continued until 1850.

The cultural revolution did not confine itself to the book trade nor to the meteoric rise of the novel up the literary generic hierarchy, moreover. Chittick explains that 'the democratization of politics was not only reported but reflected in the press'.[5] Important in this process was the dramatic proliferation in the number of cheaper journals in circulation (many with intellectual aspirations), the advent of 'penny dreadfuls', cheap weeklies and (by the end of the century) comics. As consumers of culture were afforded greater access and choice in the cultural market-place, it does not necessarily follow that all cultural products benefitted similarly from this market expansion. This is particularly true for many Victorian poets, of course, whose specialized craft suffered under new competition from seemingly more accessible media. Isobel Armstrong's *Victorian Poetry* is a superb account of the Victorian poet's 'modern' consciousness of his/her 'secondary' role in a culture which was no longer the undisputed preserve of the élite: 'To be modern', as Armstrong puts it, 'was to be overwhelmingly secondary'.[6]

The written word was not the only medium whose distribution and function was subject to metamorphosis. The 1843 Theatre Regulation Act effectively conceded that not even the rule of law could prevent the working classes from exercising a stake in the theatre. The Licensing Act of 1737 had attempted to improve the quality of British drama (and exert social control) by cutting it off from its popular roots. Only the 'legitimate' theatres, Covent Garden and Drury Lane, could produce 'legitimate' drama, while the 'illegitimate' others could only produce plays accompanied by music. The most ridiculous effect of the cultural segregation and aesthetic engineering attempted was that in 1789 an actor, John Palmer, was called a 'rogue and vagrant' for speaking prose at the Royal Circus and sent to prison.[7] The most significant result was that the popularity of the legitimate theatre waned, while the import of melodrama to the illegitimate theatres from the late eighteenth century onwards appealed hugely to the artisan and working classes, who felt the genre to be their own.[8] Though much theatre criticism emphasizes the conservatism of melodrama, the very presence of the working classes in the theatres in such large numbers to watch plays which were often overtly political – particularly domestic melodrama – was a cause of real concern for elements of the political and theatrical establishments. In the end, the legitimate theatres survived by incorporating melodrama into their programmes.

Elements of twentieth-century cultural theory have taught us to be suspicious, however, of any narrative which appears to demonstrate the unambiguous triumph of populist taste; for every theatre historian who regards the history of the Victorian theatre as a radical, subversive violation of a privileged cultural space by the working classes, there will be another who regards it as a tale of cultural appropriation by the bourgeoisie. Resolution of this analytical opposition is less important for our purposes than recognition of the fact that it is in the Victorian period that the term 'popular' becomes deeply problematic: it is at this time that it becomes especially difficult to determine whether 'popular' culture is an expression of the populace or the imposition of those with economic capital. The mass production enabled by industrialization and the capitalist system of economics which strengthened this new mode of production means that, in the nineteenth and twentieth centuries, 'popular' culture is as likely to signify 'mass' or 'consumer' culture as it is likely to mean 'street' culture or the culture *belonging* to the populace. Perhaps more significant in the context of this introduction is the fact that if there was uncertainty over the meaning of 'popular culture', there was related uncertainty over the very definition of culture. The Victorian period should thus be crucially important in cultural history and theory because it 'begins to conceptualise the idea of culture as a category'.[9]

The advent of consumer culture and its divorce from minority 'high' culture is often taken as a feature of 'modernity'. There is inevitable disagreement about the beginnings of mass culture; whereas this introduction has emphasized the importance of the early nineteenth century, the late nineteenth century is also a 'popular' landmark for cultural theorists,[10] and the global popularity of radio, television and film in our own century has given new meaning to the word 'mass'. The concept of 'culture' as it is now debated has of course metamorphosed with the radical changes to systems of communication brought about by the technological revolution of the twentieth century. But it is also true to say that the technological grew out of the industrial revolution – and that both are creatures of capitalist economics.

The search for fixed origins is without doubt an illusory one and certainly too lengthy to address satisfactorily in a brief introduction. In institutionalized academia, origins and definitions can be determined by convenience and vested interests and this introduction is no different in this respect. The above examples of cultural shifts in the Victorian period are all taken from our own area of interest, the main genres of English Literature, and make no claim to represent a panoramic sketch of the entire cultural landscape of the period. The discipline of Cultural Studies is contemporary in its main focus and its leftist roots mean that 'popular' culture is the cultural site of most interest; alternatively, the traditional Arnoldian study of English privileges the 'high' culture of the past. This volume will demonstrate the continuities between the past and the present, the high and the low. *Rethinking Victorian Culture* is predicated on the assumption that the history of modern culture is paradoxically Victorian – and vice versa.

The project of this collection has another kind of appropriateness, moreover. Many of the contributors are insitutionally defined as 'English Literature' academics, and the anthology views culture, for the most part, from the literary end of the academic telescope, making few claims for itself as an extra-literary cultural critique. However, what these essays will show is that Literary and Cultural Studies are not necessarily the oppositional, antagonistic disciplines caricatured above and are moving perhaps ever closer. That the price of such a dissolution of institutionalized categories is the difficulty of clearly delineating between disciplines is perhaps a price worth paying. It is also fitting that two disciplines sharing a common lineage should benefit from the recognition of what they have in common, as well as an informed understanding of – dare we say it – 'cultural' differences.

We are referring here to the fact that Matthew Arnold, the major cultural theorist of his age, has often been viewed as one of the founders of the academic discipline of English Literature as it was conceived in Britain until literary theory metamorphosed its rationale. Arnold was the first Professor of Poetry at the University of Oxford; his conception of literature as an instrument of individual perfectibility which enabled its possessor to civilize self and society was hugely influential in English Studies right up until the work of his disciple, F. R. Leavis, a century later. Leavis's distinct brand of English helped to ensure that the subject was regarded with due seriousness and hence aided its institutional survival by defining 'Literature' in opposition to popular culture. His rationale was nothing if not overt. In *Mass Civilization and Minority Culture*, Leavis stated:

> In any period it is upon a very small minority that the discerning appreciation of art and litearure depends: it is [...] only a few who are capable of unprompted, first-hand judgment. They are still a small minority, though a larger one, who are capable of endorsing such first-hand judgement by genuine personal response [....] The minority capable not only of appreciating Dante, Shakespeare,

Baudelaire, Hardy (to take major instances) but of recognising their latest successors constitute the consciousness of the race (or a branch of it) at a given time [...] upon this minority depends the power of profiting by the finest human experience of the past; they keep alive the subtlest and most perishable parts of tradition. Upon them depend the implicit standards that order the finer living of an age, the sense that it is worth more than that [...]. In their keeping [...] is the language, the changing idiom upon which fine living depends, and without which distinction of spirit is thwarted and incoherent. By 'culture' I mean the use of such language.[11]

Cultural Studies, on the other hand, regards Williams's *Culture and Society* as one of its founding texts;[12] Williams was also a Professor of English Literature (at the University of Cambridge), whose own work developed in opposition to the élitism he perceived in the Leavisite model of English. English and Cultural Studies have thus from their birth been intimate, though not always harmonious, relations – and the Victorian period has been formative in the history of both disciplines. While literature was usually studied in cultural 'context' from the writings of Arnold to those of Leavis, priority was unambiguously given to 'Literature' (with a capital 'L'). The New Critics' insistence on the divorce of text from context represents the separation of Literary from Cultural Studies at its most complete. Literary theories which developed partly as a critique of New Criticism and the teachings of Leavis have been largely responsible for the cross-fertilization of Literary and Cultural Studies, providing a common critical landscape, if not a common rationale for both disciplines. Even if one relates the emergence of literary theory to the writings of the linguist Ferdinand de Saussure, cultural theory also regards Saussure (via structuralism) as a founding influence.

It would be misleading to suggest that English and Cultural Studies are two sides of the same coin, however, as the academic study of culture is not co-extensive with the study of language and literature (Sociology, for example, has been particularly formative in the development of Cultural Studies). The methodologies employed in English and Cultural Studies, moreover, can be (but are not necessarily) distinct. What the emergence of 'Cultural Studies' as an academic discipline in its own right seems to have achieved is a detailing and a questioning of the broad outlines of the work of Williams in particular. In its turn, the study of English has had to respond to the challenge of Cultural Studies constructively or, in some cases, combatively.

The diversity of essays in this volume would perhaps not confirm Arnold's hopes for 'the study of perfection' and for the role of the intellectual in forging an organic, civilized culture; with his hopes for a 'common culture', Williams might be equally disappointed. What all the essays in this collection have in common, however, is an implicit belief first that it is helpful to study Victorian literature in conjunction with both modern and Victorian culture, and second that an understanding of Victorian culture is crucial to an understanding of the concept of culture *per se*. In the Victorian period, as Isobel Armstrong maintains, 'since the very notion of a culture was new, and the idea of the minority intellectual, this entailed constructing the idea of culture and defining what in particular a literary culture was'; this collection presents itself as a contribution to the ongoing construction of the idea of culture.[13] We hope that the anthology will help to redress both the tendency of Cultural Studies to focus on the contemporary period and the persistent (though by no means universal) tendencies within Literary Studies to regard culture as simply 'background' or 'context'. The eclecticism of the essays in this volume is no doubt open to interpretation as a celebration of the multiplicity particularly valued in postmodern theory and in Cultural Studies. But as the centenary of Victoria's death approaches, our aim is that the voices in this volume will contribute positively to ongoing, necessary dialogue between the studies of literature, culture and the Victorian period.

Stefan Collini's essay offers a critique of the claim that 'the story traced in Williams's *Culture and Society* should be recognized as the founding history or legitimating genealogy of contemporary cultural criticism'. He explores the distorting binary oppositions which have, since the Victorian period itself, dominated the intellectual historiography of the nineteenth century. Among these he discusses the division between writers thought to emphasize warm feeling and those emphasizing cold rationality, the Williams-inspired opposition of 'culture' and 'society', and the institutional or pedagogical categories of 'non-fiction prose' and 'cultural criticism'. He compares the New Critical concern with form over content to a more recent attempt to read Arnold, Ruskin, Carlyle and others as either unceasingly politically radical or unswervingly politically conservative, and finds both approaches inadequate, proposing instead greater attention to the question of why these texts have generated such intensely polarized responses.

John Lucas questions the validity of the term 'Victorian' as a designation for a period which, as he shows by reference to fiction, poetry and reportage of the late nineteenth century, was often hostile to the figure of the queen and to the institution of the monarchy itself. Lucas emphasizes the importance of the impact of the Paris Commune on English Republican sympathies; he later moves forward in the century to discuss literary responses to the celebrations of the queen's Golden Jubilee. To refer to the period as a whole under the term 'Victorian' is, Lucas argues, to exclude from literary and cultural history the strong forces of anti-monarchism which his essay documents.

Kate Flint's '"The Mote Within the Eye": Dust and Victorian Vision' offers a panoramic cultural history of dust in the period, analysing its importance in crucial intellectual debates about perception and the material world, as well as its symbolic significance in literary texts. The visible physical reality of dust and its invisible, imaginative possibilities ensure its importance to Victorian culture in its widest sense – to citizens, scientists, theologians, philosophers, art historians, sociologists and creative writers – and to theories of subjectivity.

Stephen Prickett's essay examines the role played by the developing science of ethnography in Victorian criticism of the Bible and of Christianity's role in culture. Focusing on Arnold's *Culture and Anarchy* and Kingsley's novel *Hypatia*, among other texts, Prickett connects the nineteenth-century reification of German Protestant culture with an attempt to develop a form of Christianity which could be presented as Aryan rather than Semitic. Prickett's essay emphasizes at the same time the international influences on Victorian religious and scientific culture, and that culture's attempts to excise the possible influence of non-Indo-European races.

In his essay on 'Dickens and the Millennium', Andrew Sanders examines the milleniarists of the 1830s and 1840s, a largely overlooked feature of early nineteenth-century popular culture. He reconsiders millenarianism in relation to the writings of Dickens, concluding that Dickens was the first truly 'modern' writer. Brian Maidment's essay is also concerned with the consumption of time; Maidment's argument centres on the 1830s and 1840s and examines year-books and almanacs as part of an attempt to regulate temporality. Maidment shows how almanacs (which despite being a very popular medium have received little critical attention) negotiated between ecclesiastical and agricultural conceptions of time in their attempt to represent the past and predict the future. Maidment deals with the development of official temporality as a means of disciplining society, and considers the implications of this new temporality for popular Victorian culture.

John Coyle and Richard Cronin's essay sees the poetry of Tennyson's Cambridge years as typical of his desire to situate himself and his poems on the boundary between public and private spheres. They argue that Tennyson sought to inhabit a shadowy borderland without committing himself to any pre-set social, political or artistic pattern. Their essay juxtaposes Tennyson's personal psychological situation with the attack on the class values and religious hegemony of the Church of England that were embodied in Oxford and Cambridge of the early Victorian period, and argues that Tennyson's centrality in the poetry of his age was both partly formed by and formative of the Victorian conception of the poet as authoritative Other.

Juliet John analyses the so-called Newgate debate of the 1830s and 1840s, arguing that the controversy surrounding the crime fiction of Dickens, Edward Bulwer and William Harrison Ainsworth was caused by anxieties about the power of popular culture at the dawn of a 'modern' age. Her discussion of *Oliver Twist* focuses on the self-reflexive treatment of Newgate fiction and melodrama, arguing that in this early novel Dickens is interrogating the power of specifically popular cultural modes. John reads Dickens as utilizing and exploring the potential of emotion and pleasure – the generation of which is so crucial to the appeal of popular culture – as ideological vehicles. She appeals for more analysis of the function of emotion and pleasure in the power dynamics of an emergent Victorian mass culture.

Francis O'Gorman's essay argues that Victorian natural history, as a form of knowledge more accessible in the culture than the increasingly professionalized languages of empirical science, deserves more prominence in accounts of the formation and transmission of ideas about nature's value and human beings' relationship with it. The essay examines the constitutive discourses of a particular and popular example of such writing: Charles Kingsley's *Glaucus*. It analyses the text's claim that natural history is morally improving, and examines the discourse of chivalry appropriated to express that claim. O'Gorman discusses aspects of the text's construction of the natural world in theological, political and aesthetic terms, and looks particularly at its deployment of imperial discourses in configuring the naturalist's relationship with nature. His argument locates *Glaucus*'s attitude to the morality of specimen collecting within the wider context of moral debate in Victorian natural history and concludes with a discussion of natural history's place in relation to the values of professionalized empirical science in the culture.

Alice Jenkins's essay deals with the autobiography of Mary Somerville. Her reading of the Victorian mathematician Somerville's account of her own life explores questions of textual mediation and the nature of 'originality'. Her essay examines Somerville's account of her own identity as a woman and as a Scot, and discusses the role of Somerville's daughter in editing the autobiography and making her mother's scientific success acceptable to Victorian standards of femininity.

Pam Morris uses her analysis of Dickens's *Our Mutual Friend* to explore some of the main ideological and social forces of the 1860s and vice versa. She emphasizes the need to complement the bold outlines of Marxist social analyses with a more detailed, specific understanding of the interaction of cultural and economic forces with the 'main currents of feeling' of the time. As theoretical tools, Morris uses Marx's *The Eighteenth Brumaire of Louis Bonaparte*, Raymond Williams's concept of 'structures of feeling', and the writings of Pierre Bourdieu in conjunction with intellectual debates of the 1860s – specifically Darwinian ideas and Max Müller's theories of language.

Anthea Trodd analyses the attempts of late-Victorian women writers to construct a history of women and women's writing around the ideas of the occult, the private and the irretrievable. Anne Thackeray Ritchie's emphasis, in *A Book of Sibyls*, on sybilline fragments as the most characteristic surviving form of women's writing, and her employment of ghostly strategies in the structure of her own literary history, prioritizes a certain kind of writing by women in order to evoke a lost world of strong feeling. Subsequent attempts to recover a women's history from the late-Victorian period until Woolf's *A Room of One's Own* also foreground a sympathy depicted as feminine in both the findings and the process of historical reconstruction.

Gail Marshall's essay takes up Trodd's interest in the impact of late-nineteenth century women artists and echoes Alice Jenkins's concern with women's autobiography. 'Actresses, Autobiography and the 1890s' combines theatre and cultural history to argue that in the 1890s, a new type of actress emerged whose story overlaps interestingly with that of the New Woman. Whereas the 'Victorian' actress had been valued primarily for her physical appeal, the 1890s actress could evade her entrapment within the visual and enhance her professional and political status by means of a variety of factors, not least her engagement with autobiography and her participation (on stage and page) in an autobiographical aesthetic.

Themes of civilization and consumerism in the political, economic and aesthetic writing of the late nineteenth century are examined in Regenia Gagnier's essay. Gagnier argues that around the mid-point of Victoria's reign a shift took place from an emphasis on labour in production and reproduction as the determinants of worth in both literature and industry, towards a belief in the consumer's tastes and desires as the crucial factors in an assessment of value. Her essay regards the late Victorian sense of its own modernity as dependent on the consumption of the past, and reads questions of class, race and gender as part of an economics of consumption which was thought to constitute 'civilization'.

Notes

We are grateful to Routledge for allowing us to reproduce above a short extract from the introduction to Juliet John (ed.), *Cult Criminals: the Newgate Novels*, 6 vols (London: Routledge, 1998), I, v–lxxi.

1. Raymond Williams, *Culture and Society, 1780–1950* (1958) (London: Chatto & Windus, 1967), pp. 332–8.
2. See Matthew Arnold, *Culture and Anarchy*, ed. John Dover Wilson (Cambridge: Cambridge University Press, 1932; repr. 1994), Chapter 1 ('Sweetness and Light'), pp. 43–71.
3. Michel Foucault's work sees the 'modern' period as rooted in the Enlightenment but actually arising from a self-consciously critical questioning of Enlightenment values first performed by Kant in his essay 'Was ist Auflärung?' (1784). Rejecting a positivist view of history, Foucault often refers to the modern period in necessarily vague terms as encompassing the last two centuries and originating at the turn of the eighteenth and nineteenth centuries. He is more precise when dating 'modernity'; Kant's 'Was ist Auflärung?' is for Foucault 'a point of departure: the outline of what one might call the attitude of modernity'; 'modernity', he defines 'rather as an attitude than as a period of history', an attitude encapsulated by the nineteenth-century writer, Baudelaire. See Foucault, 'What is Enlightenment?', in *The Foucault Reader*, ed. Paul Rabinow (Harmondsworth: Penguin, 1984), pp. 32–50 (pp. 38–9).

    Like Foucault, Isobel Armstrong identifies the 'modern' period with the emergence of a particular attitude rather than with specific historical events (although inevitably the two are related). For Armstrong, Victorian poets were the first writers to think of themselves as 'modern' and 'to be "new", or "modern" or "post-Romantic" was to confront and self-consciously to conceptualise *as* new elements that are still perceived as constitutive forms of our own condition' – *Victorian Poetry: Poetry, Poetics and Politics* (London: Routledge, 1993), p. 3.

    Arguments surrounding definitions of 'modernity' and 'the modern' are of course endless and we hope that some of the essays in this collection will extend the debate beyond the necessarily brief confines of this introduction (see, for example, essays by Sanders, John and Gagnier).
4. Elliott Engel and Margaret F. King, *The Victorian Novel before Victoria: British Fiction during the Reign of William IV, 1830–37* (London: Macmillan, 1984), p. 5.

5. Kathryn Chittick, *Dickens and the 1830s* (Cambridge: Cambridge University Press, 1990), p. 24.
6. *Victorian Poetry*, p. 3.
7. Joseph Donohue, *Theatre in the Age of Kean* (Oxford: Blackwell, 1975), p. 49.
8. See Michael Booth, 'Melodrama and the Working-Class', in *Dramatic Dickens*, ed. Carol Hanbery MacKay (Basingstoke: Macmillan, 1989), pp. 96–109.
9. Armstrong, *Victorian Poetry*, p. 3; Armstrong is in fact talking specifically here about Victorian poetics.
10. See Antony Easthope, *Literary into Cultural Studies* (London: Routledge, 1991), pp. 76–7; Easthope argues that 'In the West from at least the 1890s popular culture and its distance from high culture became largely incorporated into the great contemporary media [...] whose condition of existence was modern technology' (p. 76), and offers some interesting statistics about the expansion of the mass media from the 1890s onwards.
11. Minority Pamphlets, I (Cambridge: Gordon Fraser, 1930), pp. 3–5.
12. The work of Richard Hoggart and Stuart Hall, of course, has also been formative in the development of Cultural Studies.
13. *Victorian Poetry*, p. 27.

2
From 'Non-Fiction Prose' to 'Cultural Criticism': Genre and Disciplinarity in Victorian Studies
Stefan Collini

1.

I take my text from a source which, if not quite biblical, may be regarded by some as having a roughly comparable authority. It comes from the book which the MLA, under whose aegis it was produced, obviously intended to be what it would call a 'landmark' volume: *Redrawing the Boundaries: The Transformation of English and American Literary Studies*, edited by Stephen Greenblatt and Giles Gunn, published in 1992. Each chapter in this book offers itself as an authoritative survey of recent developments in each of the major fields of contemporary literary studies, some defined in terms of periods, such as 'Mediaeval' or 'Romantic', some in terms of critical approaches, such as 'Psychoanalytic Criticism' or 'Gender Criticism'. Among the latter is a chapter simply entitled 'Cultural Criticism', jointly written by Gerald Graff and Bruce Robbins. Graff and Robbins introduce their chapter with a discussion of Raymond Williams's book, *Culture and Society*, first published in 1958. They describe the book as 'the story of the making of a counter-culture', and they continue: 'This story, more or less as Williams tells it, has been for us the founding story of cultural criticism'.[1] I shall leave aside for now the tricky question of what is here meant by 'cultural criticism' and how far it is, or should be, recognized as a distinct field of literary study; we may content ourselves with the rather lower-level empirical observation that there's a lot of it about. But the claim I want to concentrate on is the claim that the story traced in Williams's *Culture and Society* should be recognized as the founding history or legitimating genealogy of contemporary cultural criticism.

That this issue is of particular relevance to Victorianists is confirmed in the chapter in the same volume entitled 'Victorian Studies', by the distinguished American scholar George Levine. Levine also gives considerable prominence to *Culture and Society*, calling it 'one of the most influential books of the last half century', 'a genuine classic that remains indispensable', and he goes on to claim that 'Williams's notion of culture came to pervade the work of many Victorianists'.[2] And for enlightenment about just what this notion of 'culture' is, or has been widely understood to be, we may go back to the chapter by Graff and Robbins where they say: 'At the story's center is a concept of culture that is presumed to be "critical", an antidote for the dissociated and disembodied social actuality'.[3] Bruce Robbins repeats much the same claim in his own widely admired book, *Secular Vocations*, published in 1993, when he says the history described in Williams's classic work 'marks the founding of the cultural criticism that is the intellectual's stock-in-trade, for it produces a concept of "culture" that is "critical" – set against social actuality – *by its very definition*'.[4]

Culture and Society touches on a wide range of writers from Burke and Cobbett at the end of the eighteenth century up to Tawney and Orwell in the middle of the twentieth, but the animating focus of the book is on the major Victorian prose writers such as Carlyle, Mill, Arnold, Ruskin and Morris. This is, of course, one of the main reasons why, as Levine says, the work has been particularly influential in Victorian studies, though I think that has also been true for a reason that Levine does not mention. For the truth is that since the beginnings of the scholarly study of the Victorian period earlier this century, the writings of these figures have consistently posed problems for the prevailing schemes of academic categorization. In English departments in particular – and I shall return to why they have been the most relevant institutional setting for the story – it has long been unthinkable entirely to ignore the prose writings of figures such as Carlyle, Arnold, Ruskin, and company. But at the same time, it has been hard to know quite what to do with them, and Williams's book has provided one very attractive way of addressing them, a way that has seemed to many to have the merit of directly connecting them to one cherished justification for the study and teaching of literature.

In this essay, I want to propose that there are specific historical reasons why Victorian scholarship has been especially troubled by such apparently anomalous writings, and that the shifting treatment of them during this century has provided a particularly revealing index of the tensions and limitations in the reigning categories of literary study, including, above all, the definition of 'literature' itself. I shall begin by suggesting that this is true of what has long been the most familiar category under which to include them, namely that of 'non-fiction prose'. But I shall also argue that the current tendency to attend to them chiefly as the founders of an 'oppositional' notion of cultural criticism, in the terms laid down by Williams's classic treatment, far from resolving these difficulties, betrays equally revealing tensions in some of the most influential contemporary conceptions.

2.

Let me first give what the Michelin Guides term *un peu d'histoire*. The most relevant bit of history here is the story of the separation and institutionalization of the newly specialized academic disciplines in the late-nineteenth and early-twentieth centuries. There was from the start, and perhaps still is, a tension between seeing 'English' as one professional scholarly discipline alongside others, with its own distinctive techniques and subject-matter, and seeing it as something of a residual subject, a space which licensed continued meditation on those large questions about morality and the meaning of life which had been expelled from the increasingly specialized discourses of philosophy, history and, a little later, the social sciences. Thus, alongside the philological and literary-historical emphasis that, in the heyday of the influence of German historical scholarship, marked English's bid for disciplinary respectability, there co-existed the older notion of 'letters', which for the great reading public remained, and perhaps in some ways remains still, a capacious and hospitable category that included essays, autobiography, history, and so on. In the early university syllabuses, and still more in the major publishing ventures of this period, no very restrictive notion of 'literature' was being deployed, and since the work of the great Victorian sages or moralists was most definitely not being claimed as part of the subject-matter of the newly professionalizing disciplines of philosophy and history, it remained within the domain of 'letters'. Indeed, in such a representative handbook as Saintsbury's *A Short History of English Literature*, first published in 1898, more space is given in some chapters to the essayists and historians than to the novelists. One interesting illustration of the persistence of this extended conception well into the twentieth century is provided in a recent account of criticism since 1890, which indicates the extremely wide range of prose writers included in successive editions, right up to 1938, of Arnold Bennett's widely-selling guide entitled *Literary Taste: How to Form It; With Detailed Instructions for Collecting a Complete Library of English Literature*.[5]

Another interesting example of the persistence of these older attitudes into the era of greater specialism comes from, of all unlikely places, Cambridge in the early 1920s when the English Tripos was being given its distinctive form. There, the so-called 'critical revolution' was presided over by the exceedingly unrevolutionary figure of Sir Arthur Quiller-Couch, and in planning the new syllabus 'Q' confided to a correspondent in 1922:

> What I'm groping after is a second Part of the Tripos which shall i) mainly concern itself with English thought, and ii) be a stiff test of our men's capacity to write (which includes thinking). What we want is a Part II that will turn out men provided with some useful principles for statesmanship, the better journalism etc., and some knowledge of what Englishmen have thought from time to time.[6]

This passage obviously invites several kinds of comment, but I shall here content myself with mildly observing that it is interesting that he thought these things should be part of a degree in English. In any event, this was the direct origin of the paper called 'The English Moralists', in which, of course, Carlyle, Arnold, Ruskin and company played a prominent part. For many years thereafter the guardian of the flame of 'the English Moralists' was Mr Background himself, Basil Willey, and it is not entirely irrelevant to my theme to note that in the early 1960s it was one of the first courses for which the newly appointed Raymond Williams began to lecture.

But the main lines of the story from the 1920s and 1930s onwards are, of course, provided by the increasing dominance of the critical assumptions derived from the practice of Eliot and the New Critics, and theirs was the pedagogy that came to dominate Anglo-American literary study from the 1930s to at least the early 1970s. Within this critical practice, the almost religious significance attached to the concept of 'literature' was coming to be focused on the Holy Trinity of poetry, drama and the novel. Under this regime, the recalcitrant mixed-mode texts of the Victorian essayists and moralists did not fare well: they did not sufficiently display the transforming work of 'the imagination', and anyway they were too obviously *about* something outside themselves.[7] As John Guillory, among others, has emphasized, a defining property of the New Critical conception of 'literature' was precisely that it was non-propositional, that literature achieved its effects (and its status) by its enactment of tension, ambiguity, irony and so on, rather than by any kind of discursive statement.[8] But for the most part it still seemed unthinkable to omit the work of the Victorian moralists altogether, and so it was consigned to the dustbin category of 'non-fiction prose'. Dustbin categories have their uses, of course: they give you somewhere to hide the mess when people come to visit, and syllabuses in 'English Literature' have had good reason to worry about meeting the neatness standards of the HEFCEs of the mind.[9]

Non-fiction prose is in principle a nearly limitless category, of course: Molière's M. Jourdain should have realized that, like the rest of us, he had actually been speaking 'non-fiction prose' all his life. But in practice it has been a category which has betrayed the instabilities of the reigning definition of 'literature', in the way in which defining something negatively usually does. This puts it in such dubious company as 'non-white peoples' or 'non-proliferation treaties'; as a term, it suffers the class *ressentiment* of 'non-commissioned officers', it threatens to pay at 'non-union rates', and it has all the excitement of a 'non-sexual relationship'. It is true that the negation can occasionally be used to indicate something positive: being 'non-stop' is a good quality in trains if not in talkers, just as it's more desirable for frying pans to be 'non-stick' than Tory Home Secretaries. But in general, negative definitions produce residual categories.

The earliest recorded uses of this term are from publishers' and booksellers' catalogues in the first couple of decades of the twentieth century, and initially it had a purely functional role, though the ways in which even the most practical categories betray more general assumptions is indicated in the definition given in the 1967 edition of the *Random House Dictionary*: 'the branch of literature comprising works of narrative prose dealing with, or offering opinions or conjectures upon, facts and reality'. One way, of course, in which disciplines commonly cope with the embarrassment of past works that no longer correspond to present pedagogical practice is to assign them to the annexe of 'the history of' the discipline in question, producing categories (and courses) that function rather like a home in which to put awkward elderly relatives. It is interestingly unclear what 'the history of literary criticism' is the history *of*, so heterogeneous are the kinds of writings conventionally grouped under that label, but one or two of the more famous Victorian essays are usually included – indeed, Arnold's much-anthologized essay 'The Study of Poetry' has been treated as a (soon rejected) rough draft for CCUE's 'mission statement'.[10] However, most of the works in question by the Victorian moralists are not programmes for, and still less examples of, 'literary criticism', and yet there is an uneasy acknowledgment that their broader social criticism played a crucial part in legitimating the cultural value of literature in the first place. This produced the odd outcome that a book or essay originally written to make the case for the humanizing power of great literature could not easily be accommodated within the genres that in the course of the twentieth century became constitutive of the resulting definition of 'literature'.

Revealingly, the kinds of writings I am discussing have often tended to move unsteadily between being treated as part of 'literature' proper and being of significance to literature as 'background' – or, as we now more hermeneutically pronounce it, 'context'. Walter Pater may be 'context' if one is writing about Oscar Wilde, but in another setting he can also be 'non-fiction prose' in his own right in a way that, say, John Addington Symonds never is. John Stuart Mill, Auguste Comte and Herbert Spencer all get discussed as part of George Eliot's 'context', but of these only Mill is likely to be addressed by literary scholars in his own right. The unsteadiness of the boundary between 'non-fiction prose' and 'intellectual context' partly reflects the never entirely settled conflict between the claims of critical and historical methods in the study of English.

I should emphasize that I am here concentrating on the *classificatory* problems generated by the development and justification of academic disciplines. This obviously has not stopped a lot of good work being done on the Victorian moralists, some of it by scholars based in English departments, though even the best examples have to struggle with these general issues. For instance, in his influential and still impressive analysis, *The Victorian Sage*, John Holloway chiefly discusses Carlyle, Newman and Arnold, as well as some novelists, and he suggests that they are marked out by having 'interests of a general or speculative kind in what the world is like, where man stands in it, and how he should live', though such a description seems bound to set off an intellectual-property dispute with the philosophers.[11] And more recently, George Landow, in his *Elegant Jeremiahs*, has attempted to identify a distinct and autonomous genre of 'sage writing'. One advantage, Landow suggests, of recognizing the existence of such a genre is that 'it promises to provide a method for studying non-fiction as literature and not just as data for the study of the history of ideas'.[12] The choice of the word 'data' is interesting here, implying that historians of ideas can be expected to display all the literary sensitivity of the average social-survey questionnaire. But in that respect the phrase is emblematic of the larger issue I am discussing, namely, the unspoken assumption that if a piece of writing has what can be regarded as 'intrinsic' qualities, then it must be the property of the discipline of literary criticism.

One of the most revealing, and certainly one of the most influential, expressions of the Eng. Lit. profession's consistently frustrated desire to make Victorian non-fiction prose amenable to the reigning critical assumptions was the collection of essays edited by George Levine and William Madden in 1967 under the title *The Art of Victorian Prose*. This volume contained essays by several of the leading Victorianists of the day from both Britain and the United States, such as Geoffrey Tillotson, John Rosenberg and Dwight Culler, and the selection of the topics covered was very much a case of 'round up the usual suspects': essays on 'Macaulay's Style', 'The Use and Abuse of Carlylese', 'Style and Sensibility in Ruskin's Prose', and so on. I do not know, by the way, whether editorial restraint was exercised in the matter of titles, but it is noticeable how baldly informative they all are; though written, as we shall see, under the aegis of essentially New Critical assumptions, there is a striking absence of well-wrought urns, tangled banks, glistening crystals, and all the other sideboard kitsch of the Symbolist book title.

In their introduction, Levine and Madden claim: 'It was in the Victorian age that non-fiction became a central and dominant literary genre'. Such writing, they suggest, can no longer be treated merely as 'background': in teaching, for example, John Stuart Mill's *Autobiography*, one is discussing 'a response to experience' comparable to that found in poetry and the novel. Writing in the mid-1960s, they felt no need to make explicit the assumption that the phrase 'a response to experience' was a cue for the entry of the hero of the piece, The Literary Critic. Notice also that they begin with an example from *teaching*: the pedagogic problem posed by non-fiction prose has always been one of the chief stimuli to category-wrestling. And so, they insist programmatically that the need is for a method, 'a method or combination of methods by which non-fiction might be studied and evaluated as imaginative vision – as art'.[13]

One of the revealing things about the opening pages of their introduction is that Levine and Madden make not the slightest stab at a definition of their subject-matter: they assume that their readers will know what works are primarily being referred to as 'non-fiction'. In itself, this is, as I stated earlier, a nearly limitless category, and the implicit criteria at work emerge when they say that not enough attention has been given to 'these works', to 'what in them has caused them to survive while other prose, some of it far more popular in its own day, has been forgotten'. A few proper names are mentioned – Mill, Carlyle, Newman and Ruskin – and by the next page, a little definition by enumeration is beginning to creep in with references to 'literary criticism, autobiography, social criticism, political oratory'.[14]

They recognize that Eliot and the New Critics 'made the aesthetic attitude towards non-fiction suspect', as a result of which, they observe, 'our [notice the collusive first-person plural] instinctive critical habit of mind continues to regard the essayist, the biographer, the social critic, the philosopher as second-class citizens'.[15] There is an interesting ambivalence here in that although Levine and Madden are ostensibly urging that authors in these genres be given full citizenly status in the republic of literature, they implicitly endorse the assumption that if these writings do not display the requisite 'literary' qualities, then it's back to the 'other passports' queue. There is also a significant difference between the short illustrative list of types of non-fiction prose given here and the one I quoted above. Where the earlier list consisted of genres such as 'autobiography' and 'political oratory', the second enumerates roles that include 'the biographer' and 'the philosopher'. Illustration is not, of course, comprehensive classification, but this slight change is a passing indication of a recurring problem, namely the way these writings have become the property of other modern academic disciplines. I shall return later to the significance of the relation of 'non-fiction prose' to other *disciplines*.

In any event, Levine and Madden conclude their mildly theoretical discussion as follows: 'The correct critical attitude toward an object viewed aesthetically must then be formal rather than ethical or practical, although it may well be concerned with the ethical and practical questions which are the substance of the aesthetic object.' Given the implicit definitions with which they are working, the first part of this sentence is a tautology: to view an object aesthetically *is* to view it formally. But the repressed will return, here in the simpering guise of being 'concerned with' those troubling 'ethical and practical questions'. And it still leaves them struggling with the chief difficulty created by their own categories: 'how', as they put it, 'can one discriminate non-fictions that are intrinsically aesthetic from those that are not?'.[16] Not surprisingly, this becomes something of a heffalump hunt, but it does at least lead them into an interesting discussion of 'style'.

Having pondered what it means for a writer to have a 'style', Levine and Madden then go on: 'The very persuasiveness of a writer's rhetoric is thus carried by the comprehensiveness and self-consistency of his whole linguistic world, not simply by the cogency of his ostensible arguments'.[17] Speaking about a writer's 'whole linguistic world' seems to me a step in the right direction, not least because that will in the end have to be understood in historical rather than purely individual terms. But the passing acknowledgement of the centrality of judgements about such writers' 'persuasiveness' is surely fatal to the purity of Levine and Madden's formalism. We do indeed need to ask of a piece of social criticism or cultural analysis how far we find it persuasive, but our answer is bound to depend partly on matters which, in the aetiolated vocabulary being deployed here, we should have to term 'cognitive'. The question of persuasiveness inevitably involves measuring an author's recommendations against a picture of the relevant bit of the world which we already have: there are (if the phrase is still permissible) some facts of the matter. Moreover, assessments of persuasiveness necessarily involve *historically* informed judgements: we may find *Culture and Anarchy* less persuasive if we think that Arnold underestimated the grievances of the Dissenters or we may find *Unto This Last* (1862) unpersuasive if we think that Ruskin misunderstood political economy, but these judgements clearly involve appeal to other than formal criteria. I should add that, as so often, the individual essays in the book are in many cases discerning and valuable pieces of criticism, far better in practice than the editors' attempt to generalize their assumed common position. But that, it seems to me, illustrates the point, namely that attempts to characterize the category of 'non-fiction prose' *in general* largely reveal the tensions and limitations in the category of 'literature' being appealed to.

The conception of 'literature' which required the formation of a residual category such as 'non-fiction prose' has, of course, come under a good deal of pressure in the last couple of decades in ways that make that category seem redundant in many quarters. This is partly because post-structuralist ideas have led to a concentration on the figural nature of all writing and hence to the assertion of the equally 'fictional' status of discursive and novelistic prose. And it is partly because much recent work by literary scholars has turned its attention to the processes and constraints of cultural or ideological forces at work across the whole range of a society's activities, where so-called 'imaginative literature' does not constitute a radically different *kind* of evidence from that provided by any other complex documentary material. But intellectual change is never in practice quite so neat. Not only are older conceptions and practices still powerfully present, but the contemporary idiom often ends up reinstating the familiar categories in new terms. In some recent discussions of the Victorian period, 'non-fiction prose' has been born again as 'cultural criticism', though some of the tensions that become manifest in the use of this up-to-the-minute category do seem eerily familiar.

3.

Let me begin my discussion of cultural criticism with *un peu d'histoire* of a slightly different kind. From the earliest misrepresentations of Malthus by the 'Lake Poets', through Ruskin's travesty of political economy in *Unto This Last*, and on to Toynbee's tendentious description, looking back from the 1880s, of the long drawn-out 'argument between economists and human beings', there were figures in the nineteenth century itself who attempted to organize the cultural landscape around an essentially binary division in which everything that is selfish, mechanical and fragmented is to be found on one side, and everything that is generous, human and whole is on the other.[18] But nineteenth-century culture itself also produced its share of those who identified and repudiated the distortions which structured this binary view, and I am, of course, inclined to see in this a reinforcement of my earlier point about not reducing or underestimating the variousness and richness of the arguments, values and discursive strategies available within Victorian public debate itself. Nonetheless, for reasons that would need to be explored elsewhere, a version of this view of the great fault-line in nineteenth-century culture survived into twentieth-century thinking in two main ways: first, as part of an account of the damage done to the fabric of English society by the Industrial Revolution and its contemporary defenders, an account associated in particular, perhaps, with the work of the Hammonds; and second, as part of the demonization of an alleged Utilitarian orthodoxy by the tradition of technophobic literary-social criticism associated above all with the work of Leavis. The understanding of eighteenth- and nineteenth-century economic and cultural history displayed by both these styles of thought has, from the outset, been subject to devastating scholarly criticism, but that understanding has tenacious roots in British political attitudes and class relations, and is not so easily dislodged.

One way to characterize the enduring contribution made by Williams's *Culture and Society* might be to say that it drew upon these styles of thought to tell a story about the origins of English thinking about 'culture', and that in the process it established a new orthodoxy about nineteenth-century thought and sensibility. Williams himself was conscious of his role in installing this orthodoxy: as he later said in the interviews published as *Politics and Letters*, people now think 'the tradition' had always been recognized as such, but in fact that was not true before he wrote. Only through a process of somewhat haphazard reading, he later reported, did he 'discover' the connections.[19] What enabled him to select and organize his cast was his claim, only made explicit in the book's concluding chapter, that 'The development of the idea of culture has, throughout, been a criticism of what has been called the bourgeois idea of society'.[20] Thus, the book is essentially structured around a largely implicit contrast between 'the culture and society tradition' and 'the bourgeois idea of society', and the result was to insist on a sharp binary division in English intellectual history of the period.

To see the logic of Williams's book clearly, we need in effect to work backwards. What he was in search of, in general terms, was a position, a set of concepts and values, that would enable him to envision a transformation of British society in a way which would give adequate recognition to the ideas of equality and solidarity while not abandoning the legacy of intellectual and artistic activities conventionally referred to as 'culture'. His own temperamental as well as strategic inclination was to attempt to identify and refine a tradition of thinking and feeling of which his desired position could be represented as the outcome in present circumstances. The figures he selected for inclusion in this tradition were distinguished by their having contributed something to the development of the nexus of ideas and values that, in a re-formulated version, Williams hoped could be made effective in the present as an alternative to the individualism and élitism that, in his view, characterized the dominant middle-class culture of mid-twentieth-century Britain. What his chosen figures had in common, therefore, was that they adumbrated ideas which were in some ways critical of, or suggested alternatives to, what Williams called 'the bourgeois idea of society'. Since they were essentially selected in terms of this *negative* criterion, such ideas were inevitably fairly diverse: they included ideas of the moral health of the nation, of human wholeness, of the distinctiveness of human spiritual or imaginative activities, of the precious inheritance of intellectual and artistic achievements, of the inappropriateness of quantitative or mechanical measures of human welfare, and so on. A selection of the figures who expressed some of these ideas in particularly influential or interesting ways then become 'the tradition of English social criticism'. And Williams's success in proposing these figures as 'the tradition' has had a distorting effect on our understanding of English intellectual history in this period, by representing it as essentially divided between, on the one hand, a hegemonic position made up of laissez-faire liberalism in politics, rational economic man in political economy, and the priority of individual liberty in social ethics; and, on the other hand, a tradition of protest or critique embodying a more holist, more spiritual, more 'romantic', and ultimately more generous notion of human life and social relations. One danger in trying to impose this division on nineteenth-century thought is that in the end the only people left to represent the alternative to Williams's 'tradition' would seem to be a few exaggeratedly orthodox political economists, a travesty of Victorian social thinking we have met before but which has surely long outlived its day.

Nonetheless, it is this structuring contrast that enables Bruce Robbins to say, in the passage quoted earlier, that Williams's history 'produces a concept of "culture" that is "critical" – set against social actuality – *by its very definition*'. And similarly, Graff and Robbins in their chapter refer to the part to be played by contemporary cultural criticism in what they call 'the century-long battle' over the extent to which the values of 'culture' 'should harmonize with or challenge the values of the larger society'.[21] There may be occasions when, for expository purposes, it is convenient to speak schematically of 'the values of the larger society', but does it ever make sense to assume that the options always reduce to those of either simply endorsing or simply rejecting such values, as opposed to say, exploring, modifying, refining, using, characterizing, extending and so on?[22] There is a fine remark of Nietzsche's about how we should not be taken in by asceticism's claim to be opposing itself to the base demands of 'nature' for, as he puts it, 'this is nature against something that is also nature'.[23] In the same spirit, I think we should recognize when dealing with the great nineteenth-century cultural critics that 'this is Victorian society against something that is also Victorian society', or in other words that the moral values, social ideals and persuasive resources they draw upon are not 'set over against social actuality', but are among the parts of that actuality.

4.

It is not simply chance that has led the issue of 'non-fiction prose' to play what has been at once an undeniably central but also peculiarly vexed role in the study of *Victorian* literature rather than that of other periods. The flourishing of this kind of writing surely required a set of historical conditions which there is no space to discuss here but which included the expansion of periodical literature, the crisis of religious faith, the claims of political and intellectual liberalism, and the character of the resulting public sphere.[24] But there is also a reason which has to do with the very question of disciplinarity itself, though to get a clear view of this we have to free ourselves from the categories imposed by *Culture and Society*. For, as I have suggested, it is too reductive to see the major Victorian moralists as chiefly concerned to articulate a critique of, and alternative to, industrialism; rather, their writings manifest, above all, an attempt to get beyond all partial, sectarian or merely specialized perspectives, to find somewhere to stand, intellectually speaking, from which the most 'general' assessments could be made. This is a self-description of which we are now highly suspicious, sensing an ideological screen behind which very particular interests must be at work. Nonetheless, in trying to characterize the distinctive forms and scope as well as the informing concerns of the writings of the Victorian moralists, it remains helpful, I am suggesting, to see those writings as a response to the pressures of various kinds of specialization, intellectual, professional, economic, and so on. But what happened as disciplinary professionalism got under way in the late nineteenth and early twentieth century was that these figures became the property of 'English' *faute de mieux*, and so some of the tensions I have identified may be seen as the attempt of one increasingly specialized discourse to accommodate a body of writing whose constitutive principle was the repudiation of specialism.

Ultimately, one of the chief reasons why the great Victorian moralists continue to trouble the boundaries of the modern discipline of 'English' is because their writings attempted to fulfil the role that in the twentieth century we have come to describe as that of 'the intellectual', a role partly constituted by the attempt to find the most general perspective from which to assess particular issues. All attempts to encompass their work in a single disciplinary net are bound to encounter problems because the genres in which they wrote are inherently resistant to the notion of disciplinary specialism.

In general, the recent work in this area that has done most to help correct these various forms of one-sidedness has involved close attention to the texture of Victorian intellectual life and public debate, to forms of publication, to relations with audiences, and so on. Some of this work has grown out of concerns traditionally associated with

English departments, such as the study of Victorian periodicals,[25] but some has been informed by a wider range of historical scholarship. Without wishing to set up a new *Überwissenschaft* or to propose a panacea for wider cultural ailments, I certainly believe that a flexible and eclectic kind of intellectual history can help to remedy some of the distortions I have been discussing. It may be particularly appropriate in this case because intellectual history itself, in the sense in which I understand the term, may be regarded as having a kind of 'anti-specialist' identity. In seeking to explore and re-describe the manifold ways in which particular ideas arose out of and participated in wider cultural conversations, it inevitably trespasses on domains usually restricted to the 'pre-history' of more than one discipline. Similarly, in seeking the fullest and most illuminating characterization of figures and episodes, it is inevitably and quite properly eclectic in its approach, allowing no monopoly to one explanatory idiom and according no automatic priority to social circumstance, individual intention, intellectual tradition or any other single analytical dimension. But that, too, is obviously a case that would need to be argued more fully elsewhere.

For the present, let me conclude by repeating my main contention, namely that if, in dealing with the work of the Victorian moralists, 'non-fiction prose' was one procrustean bed, 'cultural criticism' is in danger of becoming another. The former made these writers an annexe to a narrowly formalist conception of 'literature', which, among other things, led to a neglect of *what* they were saying as a result of an excessive emphasis on *how* they were saying it. The very formulation of the category 'non-fiction prose' was a symptom of just those distorting pressures of disciplinary specialism that the figures placed within the category had been primarily concerned to resist. Potentially, 'cultural criticism' is a more hospitable category, emphasizing the *generality* of the object of their writings. But this risks being ahistorical in other ways, especially because it is founded upon the distorting binary antithesis between 'culture' and 'society'. Raymond Williams in effect contracted the concerns of the great Victorian moralists into what he called 'a critique of the bourgeois idea of society'; more recent scholars are similarly in danger of squeezing them out of shape by insisting they be classed as either 'upholders of mainstream attitudes' or 'inherently oppositional'.

If I am right about why these writers have proved to be so problematic for the commonly available disciplinary categories, then a century in which the dialectic between specialization and reaction-against-specialization can only assume greater prominence is unlikely to see a decline of interest in these figures. It may be that some of our current expectations about those we term 'intellectuals' will not survive long into the twenty-first century, and if that helps us move away from a simplistic distinction between being *either* 'oppositional' *or* 'mainstream', it will be no bad thing. But the appeal of those who strive to attain some kind of 'general' perspective will surely remain, and we must adapt our generic categories to their achievements rather than the other way round. The category of 'non-fiction prose' treated such figures as, in effect, novelists *manqué*, whose writings exhibited some of the kinds of tropes and figures whose full glory was only to be seen in properly 'imaginative' literature; at present, the category of 'cultural criticism' risks turning them into campus radicals *manqué*, whose interventions adumbrate a rather primitive critique of bourgeois society while failing to appreciate the more insidiously coercive power of hegemony. We surely owe them something better than either.

Notes

1. Gerald Graff and Bruce Robbins, 'Cultural Criticism', in *Redrawing the Boundaries: the Transformation of English and American Literary Studies*, eds Stephen Greenblatt and Giles Gunn (New York: Modern Language Association, 1992), p. 422.
2. George Levine, 'Victorian Studies', in Greenblatt and Gunn (eds), *Redrawing the Boundaries*, pp. 136, 153, 137.
3. Graff and Robbins, 'Cultural Criticism', in Greenblatt and Gunn (eds), *Redrawing the Boundaries*, p. 422.
4. Bruce Robbins, *Secular Vocations: Intellectuals, Professionalism, Culture* (London: Verso, 1993), p. 60.
5. Chris Baldick, *Criticism and Literary Theory, 1890 to the Present* (New York: Longman, 1996), pp. 56-7, 109. A revised edition of Bennett's book was published in 1938, on which Baldick comments: 'This was almost the last gasp of the extensive popular canon, soon to be eclipsed by the restrictive model of the academic–modernist syllabus' (p. 109).
6. Quoted in F. Britain, *Arthur Quiller-Couch: a Biographical Study of Q* (New York: Macmillan, 1948), p. 101.
7. As Levine observes, during the 1950s and 1960s Victorian literature 'seemed in the hierarchy of periods to be least satisfying formally at a moment of emphatic critical formalism'; 'Victorian Studies', in Greenblatt and Gunn (eds), *Redrawing the Boundaries*, p. 133.
8. John Guillory, *Cultural Capital: the Problem of Literary Canon Formation* (Chicago: Chicago University Press, 1993), Chapter 4.
9. HEFCE is the Higher Education Funding Council for England.
10. CCUE is the Council for College and University English.

11. John Holloway, *The Victorian Sage: Studies in Argument* (London: Macmillan, 1953), p. 1.
12. George Landow, *Elegant Jeremiahs: the Sage from Carlyle to Mailer* (Ithaca: Cornell University Press, 1986), p.22.
13. George Levine and William Madden (eds), *The Art of Victorian Prose* (New York: Oxford University Press, 1968), pp. xiii, vii.
14. *Ibid.*, pp. vii, viii.
15. *Ibid.*, p. xi.
16. *Ibid.*, pp. xii, xiii.
17. *Ibid.*, p. xvi.
18. For a subtle account of the origins of, and the historical distortions involved in, this remarkably long-lived piece of high-handedness, one can now turn to the third part of Donald Winch's recently published *Riches and Poverty: an Intellectual History of Political Economy in Britain, 1750–1834* (Cambridge: Cambridge University Press, 1996).
19. Raymond Williams, *Politics and Letters: Interviews with 'New Left Review'* (London: New Left Books, 1979), p. 99.
20. Raymond Williams, *Culture and Society, 1780–1950* (London: Chatto, 1958; repr. 1983), p. 328.
21. Graff and Robbins, 'Cultural Criticism', in Greenblatt and Gunn (eds), *Redrawing the Boundaries*, p. 422.
22. Similarly, there is surely culpable over-simplification in saying, as Giles Gunn does elsewhere: 'The Romantics used the term "culture" as a weapon in the cause of social and political change; the Victorians turned it into an emblem of social and political reaction' – Giles Gunn, *The Culture of Criticism and the Criticism of Culture* (New York: Oxford University Press, 1987), p. 9.
23. Friedrich Nietzsche, *The Will To Power*, trans. Walter Kaufmann and R. J. Hollingdale (New York: Vintage, 1968), para. 228.
24. I have briefly discussed some of these conditions in my book *Public Moralists: Political Thought and Intellectual Life in Britain 1850–1930* (Oxford: Oxford University Press, 1991).
25. This is now a large body of work, much of it stimulated by the *Wellesley Index to Victorian Periodicals*; for representative examples, see *The Victorian Periodical Press: Samplings and Soundings*, eds Joanne Shattock and Michael Wolff (Toronto: Toronto University Press, 1982), and Laurel Brake, *Subjugated Knowledge: Journalism, Gender and Literature in the Nineteenth Century* (London: Macmillan, 1994).

3
Republican versus Victorian: Radical Writing in the Later Years of the Nineteenth Century

John Lucas

There is a strong case for arguing that, except in the most rigorously controlled of contexts, 'Victorian' and 'Victorianism' are terms we could well do without. They are all too frequently employed in ways that are chronologically indefensible, historically dubious, intellectually confusing and ideologically unacceptable – at least, if you're a socialist. 'Victorian' in particular is used to imply a cultural and political homogeneity which, the evidence suggests, never existed. I have set out elsewhere my arguments against the habitually slack or incurious use of the term.[1] Here, I will merely note that what at any given moment seems permanent and immutable as a term for the convenient periodization of history (or literary history) doesn't have to be so. Two examples come to mind. What used to be called the English Renaissance is now regularly referred to as the Early Modern Period, and those few scholars who study the eighteenth century seldom pin to it the label 'Augustan', which was how it was recognized by an earlier generation.

It can of course be claimed that true 'Victorianism' doesn't begin with Victoria's accession to the throne in 1837 but in the latter half of the 1870s, for it was in those years that she finally won acceptance and approval. David Cannadine has famously argued that after Albert's death in 1861 and Victoria's withdrawal from public life, the nation became increasingly dissatisfied with her, that the most obvious expression of this was the formation of a number of Republican clubs in the early 1870s, and that it took a combination of chivvying and exhortation from Gladstone and Disraeli to make her see that her position was less secure than she had assumed.[2] In *Before the Socialists: Studies in Labour and Politics, 1861 to 1881*, Royden Harrison remarks that 'the story of proletarian Republicanism begins in the Autumn of 1869 with the foundation of the Land and Labour League', and he adds that the first plank in the League's platform was the nationalization of the land.[3] This is no doubt so, but it is then important to remark that the League can be legitimately linked to a continuing tradition of Republicanism, an earlier key moment of which was the middle-class radical Republicans grouped around W. J. Fox who, in the 1830s and 1840s, had so powerfully argued against the retention of the monarchy on the grounds that, among other things, it lent legitimacy to the amassing of vast properties, of wealth and of land. Behind the Foxites are the Spencean radicals of the late eighteenth century, who had argued passionately for the land to be held in common. From the Spenceans we reach back to the most radical groups of the Commonwealth period (Diggers, Levellers and so on); and behind them are John Ball and his companions with their unanswerable question, 'When Adam delved and Eve span/ Who was then the gentleman?'.

To note the continuing vitality of this tradition, now flaring up, now going underground, helps us to understand why Victoria's ascension to the throne should have been accompanied by some very determined efforts to put her at the centre of affairs. For those who championed the monarchy were uncomfortably aware that plenty of Britons thought a republic would be a preferable alternative. In this context, therefore, we should realize that the use of the railway to make possible royal progresses across the island, the restoring and enlarging of royal residences, the naming of the State of Victoria in 1851, of the Falls in 1855, and, a year later, the institution of the Victoria Cross (soldiers could now die not merely die for their nation but for the queen – most crosses were awarded posthumously), let alone royal portraiture and the dissemination of her head on stamps and coins, all add up to a sustained attempt to produce Victoria as unopposable if only because unavoidable. The apotheosis of the queen as the mother of her people and empress of a mighty Empire had to wait until the Golden Jubilee celebrations of 1887, but by then the hard work had all been done. The monarchy was safe, was, indeed, exalted. The nation had become 'Victorian'.

But had it? If we are to answer this question we need to pose two further ones. First, why did the Republican clubs emerge when they did? Second, did Republicanism wane with the clubs' disappearance by the end of the 1870s – and there is no doubt they did by and large wither and die – or has Republicanism simply been written out of the story? Cannadine implies that the Republican clubs came into existence because of dissatisfaction with the absentee queen. Harrison has a more interesting explanation. He notes that as soon as the Land and Labour League was born, 'Marx staked his claim to paternity'. The League, Marx told Engels, was 'directly inspired' by the International and its inception signals the fact that 'the workers' party dissociates itself completely from the bourgeoisie'.[4] Perhaps. Yet we should also note that the end of the 1860s saw the emergence of a newly strong trade-union movement. What is commonly agreed by Labour historians to be the first ever Trades Union Congress took place over the Whit weekend of 1868, and in August of the following year a larger, more confident Congress gathered at Birmingham. This is not to say that the Congress was awash with radical Republicans; it is rather to suggest that in such a climate of renewed radical energies, Republicanism was certain to be part of the weather.

Here, though, we come to a feature unique to the Republican moment with which I am concerned. On all previous occasions when Republicanism had become a perceived threat to the well-being of the monarchical state, the defenders of that state enlisted Francophobia to unsettle their opponents. As Linda Colley rightly notes in her *Britons: The Forging of the Nation, 1707–1837*, fear of the French proved an excellent tool for welding constituent parts of the nation into a whole.[5] In 1859, for example, some time after Colley thinks the nation had achieved a sense of untroubled identity, Francophobia was wheeled out. In the invasion scare of that year Tennyson wrote the poem 'Riflemen, Form!' (published in *The Times* on 9 May 1859 as 'The War' with the refrain 'Form, Form, Riflemen, Form! / Ready, be ready to meet the storm! / Riflemen, Riflemen, Riflemen, Form!'). Victoria reviewed the Scottish volunteers in Holyrood Park, Edinburgh, where they had assembled in order to vow their readiness to fight the French whenever and wherever the expected invasion should occur, on which occasion an eyewitness reported that a crowd of a hundred thousand 'made the welkin ring with their reiterated cheers'.[6] Nor should we forget A *Tale of Two Cities* (1859). Francophobia was, as it still is, a way of proclaiming Britishness. (Or, I suspect, 'Englishness'.)

But then came the Franco-Prussian war, its appalling aftermath; and suddenly an uncomplicated Francophobia was impossible. People found themselves on the side of the Commune and against the *ancien régime* in whose name the killing of thousands of Communards was ordered and accomplished. In other words, the British were by and large in sympathy with Republicans. We can see how troubling the events of those years were if we consider how completely Matthew Arnold moved from the position he took up in *Culture and Anarchy* (1869) to the one he adopted in the aftermath of the Commune.

Culture and Anarchy was addressed to the new mass-democratic nation, as Arnold saw the United Kingdom becoming after the passing of the second Reform Act of 1867. And for Arnold democracy could easily spell anarchy or 'Doing As One Likes' – the title of a key chapter in *Culture and Anarchy* which had been prompted by a march of working men who, unable to get into Hyde Park where they intended holding a meeting, ripped up the park's railings. What did such action portend?

> Our masses are quite as raw and uncultivated as the French; and so far from their having the idea of public duty and of discipline, superior to the individual's self-will, brought to their mind by a universal obligation of military service, such as that of conscription – so far from their having this, the very idea of a conscription is so at variance with our English notion of the prime right and blessedness of doing as one likes, that I remember the manager of the Clay Cross works in Derbyshire told me during the Crimean War, when our want of soldiers was much felt and some people were talking of a conscription, that sooner than submit to a conscription the population of that district would flee to the mines, and lead a sort of Robin Hood existence under ground.[7]

Good for them, I want to say, although I doubt that a conscription would have been so hard to raise as Arnold, for the purposes of his argument, makes out. Admittedly the enemy in the Crimean War was not France, but Arnold's appeal to a national spirit has about it an element of the xenophobia which was most often turned against the French.

But then came the events of 1870–1. They shook Arnold as they shook others, and he now finds himself making common cause with the populace, that uncultivated mass of people he had in so lordly a manner lectured in *Culture and Anarchy* for its lack of consideration for the needs of state. Arnold did not want the French upper class to emerge victorious from the post-Commune period. 'What is certain', he wrote, 'is that all the seriousness, clear-mindedness, and settled purpose is hitherto on the side of the Reds'. Seriousness, clear-mindedness, settled purpose: why, they are the very desiderata of Arnold's concept of Culture. As for the French working-man, no self-interest there. 'It is [his] clearly putting his resolve before himself and acting upon it, while the working man elsewhere is in a haze about it, that makes France such a focus for the revolutionists of Europe'.[8]

Should we include Hopkins among the revolutionists? He himself was inclined to do so. In the famous 'red' letter which he sent to his friend Robert Bridges on 2 August 1871 – it was to cost him Bridges' friendship for some 13 years – Hopkins confessed: 'I must tell you I am always thinking of the Communist future'. And then he adds,

> horrible to say, I am in a manner a Communist [...]. It is a dreadful thing for the greatest and most necessary part of a very rich nation to live a hard life without dignity, knowledge, comforts, delight, or hopes in the midst of plenty – which plenty they make. They profess that they do not care what they wreck and burn, the old civilisation and order must be destroyed. This is a dreadful look out but what has the old civilisation done for them?[9]

Hopkins has England especially in mind but his letter is assuredly written in the light of unfolding events in France. After the humiliating defeat of the French army at Sedan, a republic had been declared in Paris against the Emperor Louis Napoleon. The army was instructed to put it down. As a result at least 20,000 Communards were slaughtered by French troops still loyal to those who had sent them into a hopeless war (Communards claimed with some justification a far higher figure of 36,000, many executed after summary court-martial). With the suppression of the Commune a new government was proclaimed, ostensibly Republican but in fact dominated by monarchists. As with the Bourbons, it might seem, nothing had been learnt, nothing forgotten. Or, in words used by George Meredith's friend, F. A. Maxse, in a published lecture of 1872, 'The Causes of Social Revolt', there was a

> terrible week in May, when an ignorant soldiery entered Paris, on behalf of religion and order, and rioted in the bloodshed of Paris workers and their families, while the clerical journals hounded them on to massacre; and one of them indignantly demanded why only 40,000 Communists had been killed.

'We may dragoon, sabre and shoot down democracy,' Maxse adds. Nevertheless, the vision of social justice means that 'men will be found, some madly but others wisely, to devote themselves to an incessant struggle for radical change'. And we should note that Meredith himself, in a series of dialogues published in *The Graphic* in December 1872 and January 1873, makes clear his support for the Communards, and his contempt for those who clamour for the return of a monarchy.[10] In this he is joined by others, among them his friend A. C. Swinburne, who in 1871, that key year, published *Songs Before Sunrise*. Here, Republicanism is of the essence.

It is therefore important to note that the collection as a whole is preceded by a dedicatory poem to Joseph Mazzini, hailed as 'the world's banner bearer, / Who shall cry the republican cry,' and that the first poem proper is called 'The Eve of Revolution'. This is in many ways a restatement of Shelley's 'Ode to Liberty' (1820), and towards its close, in stanza 26, Swinburne invokes 'Swift Revolution', which will, he says, produce Freedom for all in a 'Serene Republic of a world made white'. Later in the collection comes 'The Litany of Nations', in which the various European nation states give tongue to express their sense of national identity and voice their hopes for future good. Here is the voice of England.

> I am she that was and was not of thy chosen,
>   Free, and not free;
> She that fed thy springs, till now her springs are frozen;
>   Yet I am she.
> By the sea that clothed and sun that saw me splendid
>   And fame that crowned,
> By the song-fires and sword-fires mixed and blended
>   That robed me round;
> By the star that Milton's soul for Shelley lighted,
>   Whose rays insphere us;
> By the beacon-bright Republic far-off sighted;
>   (*Cho*) O mother hear us.[11]

That Swinburne should identify the spirit of England with the regicide Milton and king-hater Shelley makes plain that the Republic far-off sighted is indeed one which has overthrown monarchy.

It will also have overthrown religion. Atheism and the beacon-bright Republic go hand-in-hand. The concluding lines of 'Hymn to Man' announce that 'Thou art smitten, thou God, thou art smitten; thy death is upon thee, O Lord. / And the love-song of earth as thou diest resounds through the winds of her wings – / Glory to Man in the highest! For Man is the master of things'. But that is the future. For the present, as 'To Walt Whitman in America' avers, Europe and England in particular suffer from the tyranny of kingship: 'Chains are here, and a prison, / Kings, and subjects, and shame'. Freedom lies in shaking to earth like dew the fetters of a monarchical state. Hence, 'Marching Song', where Swinburne appeals to England in the following sub-Shelleyan terms:

> And thou, whom sea-walls sever
> From lands unwalled with seas,
> Wilt thou endure for ever
> O Milton's England, these?
> Thou that was his Republic, wilt thou clasp their knees?
>
> These royalties rust-eaten,
> These worm-corroded lies,
> That keep thine head storm-beaten
> And sunlike strength of eyes
> From the open heaven and air of intercepted skies;
>
> These princelings with gauze winglets,
> That buzz in the air unfurled,
> These summer-swarming kinglets,
> These thin worms crowned and curled,
> That bask and blink and warm themselves about the world;
>
> These fanged meridian vermin,
> Shrill gnats that crod the dusk,
> Night-moths whose nestling ermine
> Smells foul of mould and musk,
> Blind flesh-flies hatched by dark and hampered in their husk.

And so on for several further stanzas, all of them hammering away at the parasitic nature of the monarchy and, we have surely to assume, Victoria and her numerous off-spring, who fasten like leeches or bloodsucking insects on the state's health and wealth. ('These limbs, supine, unbuckled, / In rottenness of rest, / These sleepy lips blood-suckled / And satiate at thy breast', as a later stanza puts the matter.)

Songs Before Sunrise is very much of its moment. The critic who undertook to write about the collection in the *Saturday Review* unsurprisingly commented that 'Much as [Swinburne] delights in what sued in our younger days to be called blasphemy, he delights still more, if that were possible, in the reddest or red Republicanism'. And the *Edinburgh Review* of July 1871 noted that while Swinburne dreamt of founding 'the Red Republic [...] the condition of France, and especially of Paris, during the last three months and at the present time, is the best possible commentary on the political principles more obscurely enunciated in *Songs Before Sunrise*'.[12] Both reviewers, hostile though they are, allow us to recognize that Marx was wrong to think the Land and Labour League's proletarian radicalism distinguished it from more temperate forms adopted by the bourgeoisie in the late 1860s and early 1870s. Or rather, it can be said that 'the reddest of Red Republicanism' appealed to radicals across a wide range of the social spectrum.

It also lasted longer into the 1870s than either Harrison or Cannadine assume. I have on my shelves a decidedly fugitive play printed in London and called *Edward the Seventh: On The Past and Present Times with a View to the Future*. This, together with a 'Key' to the characters, said to be produced 'By One Behind the Scenes', bears the legend, 'Published for the Proprietors, at 40, Bedford Street, Covent Garden, 1876', and its 84 pages of text are amply buttressed by pages of advertising, suggesting that the publishers expected it to sell in large quantities. The play is itself an odd affair, in which parodic echoes of Shakespeare's histories alternate with ribald scenes between Victoria and Prince 'Albor'; it ends with the newly crowned Edward proclaiming himself a reformed character and at the service of his people. Presumably we are to take this with a pinch of salt. The 'Key' is undoubtedly tongue-in-cheek when it describes Edward as able 'to test by the experience derived in his august mother's surroundings, what our feelings are as to the manner in which our Constitutional Monarchy should be conducted'. To gauge the true weight of this statement we need look no further than the caricature which stands as the play's headpiece. In this, a group of German junkers and Scots ghillies smile their approval as a plump Victoria whips the British lion, which cringes away from her. To the left of the picture John Bull watches in dismay while confiding to his female companion, 'What a pity, she *was* such a nice Lion but lor thats eight and thirty years ago'.

Below this clear refusal to celebrate the values of 'Victorianism' comes a Prologue in the form of a ballad. The setting for this is given as 'Off the Isle of Wight. Lat. Too great. Lon. Doubtful. Page of a Book not written up by the Quartermaster or Captain of the "Alberta" on August 18th, 1875'. The ballad tells of a collision between a sailing vessel, the

Mistletoe, and a steamer, crewed by Germans, carrying a freight described as 'the wearer of England's crown'. This is based on an actual event. The steamer was making all haste to Southampton, where the queen was to entrain on a 'special', for she was, so the ballad tells us, 'dying to see her Scots'. In the violent meeting of the two vessels the *Mistletoe* went down and several lives were lost, which 'made her Majesty's "special" late; / Though her people will gladly learn that she / Reached Scotland, at last, quite punctually'. The ballad concludes:

> The Captain died, and the mate was drown'd,
> And many days after the body was found;
> The 'crowner' sat, and the jury said
> No blame was on his Serenity's head.
> But the widows wept, and a sister's heart
> Still dared for a sister's loss to smart;
> But p'raps will teach such common clay
> Not to get in a Royal Sovereign's way.
>   Oh! the *Mistletoe* bow!

Although a specific incident undoubtedly helped prompt *Edward VII*, the play's anti-monarchical position owes much to its linking of the monarchy to power, patronage and unfair privilege. The crown has law on its side, as the punning reference to 'crowner' sufficiently indicates; and the 'Key' to the play repeatedly identifies courtiers and hangers-on in terms that align them with the repellent insects of Swinburne's 'Marching Song'. This is why it will not do to assume that the demise of Republican clubs in the later 1870s is coterminous with the disappearance of Republicanism *per se*.

 Consider, for example, the formation of the Eastern Question Association. Ever since the conclusion of the Crimean War there had been worries about the (im)balance of power that looked to exist between the Russian Empire and the declining Ottoman Empire. These worries became sharply focused in 1876 when it seemed that Disraeli was inclined to favour the Turkish cause. The Association was set up to oppose any British intervention in either nation's affairs. William Morris became a member of the Association and in November 1876 was sufficiently committed to its cause to agree to serve as the Association's treasurer. Meetings and rallies were held throughout the winter months of 1876–7 and when, as had long been feared, in late April 1877 Russia finally declared war against Turkey, the EQA redoubled its efforts to dissuade the British from involvement. Morris was by no means alone in thinking that Disraeli favoured the Turkish cause and he issued a manifesto 'To the Working-Men of England', in which he identified those who clamoured for war as

> these saviours of England's honour, these champions of Poland, these scourges of Russia's iniquities [...] greedy gamblers on the Stock Exchange, idle officers of the army and navy (poor fellows!), worn-out mockers of the Clubs, desperate purveyors of war-news for the comfortable breakfast tables of those who have nothing to lose by war, and lastly, in the place of honour, the Tory Rump, that we fools, weary of peace, reason and justice, chose at the last election to 'represent' us: and over all their captain, the ancient place-hunter [... Disraeli whose] empty heart and shifty head is compassing the stroke that will bring on our destruction perhaps, our confusion certainly.[13]

This is very like the roll-call of those who are identified as enemies of the people in the 'Key' to *Edward VII*.

A year passed and Britain continued to watch from the sidelines. But Morris was convinced that such was the influence of Disraeli over Victoria that the woman he habitually called 'the widow Guelph' or 'the empress Brown' was 'crazy for war' in spite of all denials. More meetings to oppose war involvement were held. The most dramatic such meeting was probably that of the Workmen's Neutrality Demonstration which took place on 16 January 1878, with an overflow meeting in Trafalgar Square. According to the next day's *Times* Morris ridiculed the Court and Disraeli (cries of 'No, no,' and 'Three cheers for the Queen'). The Chairman reminded Morris that 'it was undesirable in a Constitutional country in which Ministers were responsible to the people to introduce the name of the Sovereign into political discussion'. Morris replied that in that case the nation must offer 'a steady resistance' to Disraeli's shiftiness.

Morris's Republicanism inevitably accentuated his dislike of Disraeli. When he gave a lecture in the Birmingham Town Hall in 1879 on 'The Art of the People', he said 'History (so called) has remembered the kings and warriors, because they destroyed; Art has remembered the people, because they created'. This echoes a remark by W. J. Fox, who in one of his 1844 *Lectures, Chiefly for the Working Classes*, had protested that all across England statues commemorated kings and generals but not artists, poets, creators.[14] Here again, then, it is possible to recognize that Republican arguments belong to a continuum of radical thought, one of which Morris was well aware. His loathing of monarchical institutions modulates into an almost Shelleyan note of fury at the sentence of sixteen months' hard labour passed in July 1881 on Johann Most, a Viennese dissident in exile, who, after the assassination of Tsar Alexander II, had published an article which Morris described as 'a song of triumph'. (In passing it may be noted that in 1894 Morris would give evidence on behalf of an old comrade from Socialist League days, Tom Cantwell, who stood accused of soliciting the murder of members of the royal family.)[15]

Between them, *Edward VII* and the Eastern Question Association show that we cannot simply assume that from the latter half of the 1870s on the monarchy began to enjoy the nation's wholehearted support. Cannadine's argument may be substantially correct, in that we can date the rise of Victoria's general popularity from that moment, but there was still considerable resistance to her. Nor is it safe to argue that such Republicanism as survived did so among middle-class radical intellectuals, whereas the working class adopted her as the embodiment of their vision of Britain. This is part of that argument which sees working-class culture of the latter half of the nineteenth century as essentially 'the culture of consolation', in Gareth Stedman Jones's formulation.[16] The trouble with this is that so much of the evidence produced to support the argument is deeply tainted. It really will not do to cite songs like 'We're Soldiers of the Queen' and 'We don't want to fight' as demonstrating pure working-class allegiance to xenophobic way-hey the monarchy. Control of music halls had by the 1860s passed almost entirely into the hands of middle-class management who employed socially aspiring composers and lyricists to write the songs which were to be sung from their increasingly respectable stages. (For the licensing laws could be used to close down 'disreputable' halls.) Given this, songs proclaiming undying love for Queen and Country are only to be expected. But the research of Summerfield, Senelick, Bratton, and Peter Bailey ought to guard us against assuming that the music halls were therefore the true *vox pop*.[17] This isn't to say that they might not have been so, and in decidedly subversive ways at that. We don't know how these songs were sung, although we *do* know that Marie Lloyd's 'innocent' lyrics became risqué, even explicitly bawdy, through her stage performance of them, and it's at least possible that 'The Soldiers of the Queen' could be rendered in ways hardly consistent with pro-monarchical sentiment.

That particular song was written in 1881. Six years later came the Golden Jubilee celebrations which, as I have earlier remarked, established Victoria as the apotheosis of 'Victorianism'. Yet here again we should not assume that the entire nation was united in celebrating the triumph of the monarchy. Looking at the many poems published that year in praise of Victoria, I am struck, for instance, by Palgrave's 'Ode', which takes for granted that Victoria is 'England's Empress Queen',[18] as though he dare not speak for either Scotland or Wales (and certainly not for Ireland). True, this may amount to no more than that habitual English assumption that England *is* the United Kingdom, but it may equally be that Palgrave among others sensed that the din of approbation might not be sufficient to shut out dissenting voices. In this context it is worth noting that the great William McGonagall's 'Ode to the Queen on her Jubilee Year' includes the following stanzas:

> Let all hatred towards her be thrown aside
> All o'er her dominions far and wide;
> And let her subjects bear in mind,
> By God kings and queens are put in trust o'er mankind.
>
> Therefore rejoice and be glad on her Jubilee day,
> And try and make the heart of our Queen feel gay;
> Oh! try and make her happy in country and town,
> And not with Shakespeare say 'uneasy lies the head that wears the crown'.[19]

McGonagall's very lack of guile suggests that he knew of reasons for the monarchy to rest uneasy. If so, the following piece of near reportage acquires an added interest. The scene is an omnibus taking a group of people into the city to see the illuminations on Jubilee day itself, 21 June 1887:

> A woman [...] talked loudly about the procession, with special reference to a personage whom she called 'Prince of Wiles' [This may be intended merely to imitate cockney speech or to indicate a contemporary soubriquet for the heir apparent] [....] Beyond sat a workingman, overtaken with liquor, who railed vehemently at the Jubilee, and in no measured terms gave his opinion of our Sovereign Lady; the whole thing was a 'lay', an occasion for filling the Royal pocket, and it had succeeded to the tune of something like half a million of money, wheedled, most of it, from the imbecile poor. 'Shut up!' roared a loyalist, whose patience could endure no longer. 'We're not going to let a boozing blackguard like you talk in that way about 'er

Majesty!' Thereupon, retort of insult, challenge to combat, clamour from many throats, deep and shrill.

That comes from Chapter Seven of George Gissing's novel, *In the Year of Jubilee*.[20] The novel was written some time after the event itself and first published in 1894. Perhaps for this reason Gissing doesn't develop the scene on the bus: its potential for disrupting the accord of celebration is not explored. We might then wonder why he bothers to set the novel in that year at all. One answer is that the carnivalesque of Jubilee day provides Gissing with an opportunity to study its effects on various young women with which the novel is concerned. To say this is, of course, to indicate that the novel is far more of the 1890s than the previous decade. But *In the Year of Jubilee*, while undoubtedly of its decade, also allows Gissing to explore how the various roles and images of women which are apparent in the 1890s may have been fashioned by the 1880s. If the Golden Jubilee succeeded in producing Victoria as the ideal monarch-as-matriarch, mother to a large family which stretched as far as India and Africa, what did this mean for other women? Could they, *should* they, model themselves on her? Should they too be 'Victorian'?

No, is the answer. *In the Year of Jubilee* is very much a novel about the problems women face who choose to marry or not, or for whom the choice is not available. Marriage is a problem. So is childbearing. Conservative as Gissing had undoubtedly become in general outlook by the 1890s, I do not think *In the Year of Jubilee* can be called 'Victorian' and it certainly doesn't celebrate the values by then most closely associated with Victoria.

But a far more radical challenge to the values identified with the Jubilee is to be found in a novel published in 1888. *Out of Work* by 'John Law' – Margaret Harkness – opens as follows:

> It was the day after the Queen's visit to the East End. Whitechapel was gay with flags. Mile End had coloured banners, and festoons of red, yellow, green, and blue paper flowers 'all along the line'. Her Majesty had been enthusiastically welcomed by crowds of West End visitors at the London Hospital and the great breweries [....] Reporters were busy at work concocting stories of the royal progress through the East End for the Monday papers [....] No one would speak about the hisses which the denizens of the slums had mingled with faint applause as Her Majesty neared her destination; no one would hint that the crowd about the Palace of Delight had had a

sullen, ugly look which may a year or so hence prove dangerous. The ladies on their way back to the Queen's Hall, who had leant back languidly in their carriages, heedless of ragged men, hungry women, and little dirty children, the *blasé* frequenters of Hyde Park and the clubs, who had glanced carelessly at the people as they accompanied their wives and daughters to the People's Palace, would be quoted by reporters as philanthropic persons intent on ministering to the poor by the unction of their presence, and represented by the artists as so many unselfish ladies and gentlemen, who had given up an afternoon's pleasure-hunting in order to gratify the eyes of under-paid men and over-worked women by their shining hats and charming bonnets.[21]

Out of Work is among the first of those late-nineteenth-century works, fictional and non-fictional, which plunge into 'darkest London' in order to deal with the lives of the poor, the 'under-class', in determinedly truthful and hence disenchanted terms. Determination is likely to be all-important. For the naturalism which goes hand-in-hand with most of these accounts takes for granted the lives of the city poor as exemplifying a struggle for survival to which all else is sacrificed. This is not however the case with Harkness. Her socialism has no truck with the resigned pessimism of Gissing, who in such novels as *The Unclassed* (1884) and *The Nether World* (1889) saw nothing to be done, any more than did Arthur Morrison in his *Tales of Mean Streets* (1894) and *A Child of the Jago* (1897). In contrast, the sullen looks of the East End poor in *Out of Work* may, Harkness notes, 'a year or so hence prove dangerous'. And she meant it. For much of her novel explores the desperate shifts to which the under- and unemployed are put in order to find work or money. Those who think an afternoon's philanthropy and an occasional royal visit will be sufficient to reconcile the poor to their condition had better think again.

Harkness had in mind certain specific events of Jubilee year and she could be confident that her readers would be fully aware of them. For the year of the Jubilee was also the year of 'bloody Sunday', which forms a crucial episode in *Out of Work*. 'Bloody Sunday' was 13 November. In the weeks before there had been a number of marches of the unemployed through London. On this occasion, as previously, the march culminated with a meeting in Trafalgar Square. The police decided to break the meeting up, charged through the square on horseback and in the ensuing *melée* hundreds of working men were injured and three were killed. The next morning a leader in *The Times* naturally spoke up for authority in its analysis of this event.

> Putting aside mere idlers and sightseers [...] and putting aside a small band of persons with a diseased craving for notoriety [...,] the active portion of yesterday's mob was composed of all that is weakest, most worthless, and most vicious of the slums of a great city [....] [N]o honest purpose [...] animated these howling roughs, it was simply love of disorder, hope of plunder, and the revolt of dull brutality against the rule of law.²²

Harkness's sympathetic account of how sheer desperation drives some of her characters to the march and subsequent meeting effectively demolishes *The Times*'s accusations levelled against those taking part in the events which culminated in 'bloody Sunday'. Her novel is a most important intervention, a challenge to the orthodox account of monarchy as for the good of the nation. It plainly isn't any good for the thousands like Jos whose only alternative to the degradation of unredeemed poverty is protest against a system which denies them labour or hope.

But protest or no protest, Jos cannot find work. At the end of the novel, penniless and hopeless, he decides to return to the village from which he had set out for London, confident that in the city employment in abundance awaited him. On the tramp he comes across parkland which, he is assured, is 'shut to tramps and vagrants'. So much for a dream of rural freedom. Soon afterwards he passes near the walls of Windsor Castle, and a little later tumbles into a public house, where he overhears the local gamekeeper extolling the queen's virtues.

> 'Fifteen years back, when I was ill with rheumatics, she came to see me. It isn't every man as can say as the Queen's been to see him. I was in bed, and that bad in my joints I couldn't move myself. She's getting up in years, and, if she can't do what she used to do, I'm not the man to forget that she came to see me when I'd rheumatics.'
>
> 'It's nothing to do with the Queen that the land doesn't pay, and the farms is empty,' remarked the host. 'She's done her best with this Jubilee business.'
>
> Margaret Harkness, *Out of Work*, Chapter 19, p. 247

But of course that best cannot help Jos nor the many thousands like him. Predictably enough, when he finally reaches his native village there is no work for him. Soon after he is dead of cold, exhaustion, and lack of food.

Out of Work frames its tale by references to the Jubilee and Harkness throughout provides mordant evidence that the state of the nation – its illness – is causally connected to a social system which is structured on privilege, private possession, especially of the land, and contempt for the rights of all citizens.[23] The alternative to this must be a Republican vision of social equality and common ownership. That so many held to this vision in the closing decades of the nineteenth century explains why we should not be content to call the period 1870–1901 Victorian.

Notes

1. 'Love of England: Patriotism and the Making of Victorianism', in *Heart of the Heartless World: Essays in Cultural Resistance in Memory of Margot Heinemann*, eds David Margolies and Maroula Joannou (London: Pluto Press, 1995), pp. 133–47.
2. David Cannadine, 'The Context, Performance and Meaning of Ritual: The British Monarchy and the "Invention of Tradition"', in *The Invention of Tradition*, eds Eric Hobsbawm and Terence Ranger (Cambridge: Cambridge University Press, 1983), pp. 110–18.
3. Royden Harrison, *Before the Socialists: Studies in Labour and Politics, 1861 to 1881* (London: Routledge and Kegan Paul, 1965), pp. 215, 216.
4. *Ibid.*, p. 216.
5. Linda Colley, *Britons: Forging the Nation, 1707–1837* (London: Yale University Press, 1992), see esp. Chapter One, section three, 'Jerusalem the Golden'. For more on Francophobia in England in the nineteenth century, see my essay, 'Love of England', esp. pp. 139–40.
6. *Rebels and their Causes*, ed. M. Cornforth (London: Routledge and Kegan Paul, 1978), p. 125. Tennyson was not the only poet to rally to the (supposed) cause. In 1860 Martin Tupper published a ballad called 'Rule, Britannia!' in which he warned that 'France is coming, full of bluster' and urged his fellow countrymen to shoot French soldiers 'as you would a wolf!' (*Rebels and their Causes*, ed. Cornforth, p. 125).
7. The passage is to be found in the section on 'Doing as One Likes'. See *Culture and Anarchy: an Essay in Political and Social Criticism*, ed. J. Dover Wilson (Cambridge: Cambridge University Press, 1932), pp. 84–5.
8. Quoted by Lionel Trilling in his *Matthew Arnold* (London: Unwin, 1963), p. 291.
9. *Gerard Manley Hopkins: Selected Prose*, ed. Gerald Roberts (Oxford: Oxford University Press, 1980), pp. 54–5.
10. Much the most helpful account of the Meredith circle's response to the events of the Commune is to be found in Jack Lindsay's *George Meredith: His Life and Work* (London: Bodley Head, 1956), esp. pp. 188–202.

11. A. C. Swinburne, *Songs Before Sunrise* (London: Chatto and Windus, 1871), p. 79. All further references are to this edition.
12. See *Swinburne: the Critical Heritage*, ed. Clyde K. Hyder (London: Routledge and Kegan Paul, 1970), pp. 129, 136.
13. For this see Fiona MacCarthy, *William Morris: a Life for Our Time* (London: Faber, 1994), p. 382.
14. 'A Restrospect for the Year 1844' (Lecture 1), in W. J. Fox's *Lectures*, 4 vols (London: Charles Fox, 1845), I, 25.
15. MacCarthy, *William Morris*, p. 642.
16. See 'Working-Class Culture and Working-Class Politics in London, 1870–1900: Notes on the Remaking of a Working Class', in Gareth Stedman Jones, *Languages of Class: Studies in English Working Class History 1832–1982* (Cambridge: Cambridge University Press, 1983), pp. 179–238.
17. See esp. P. Summerfield, 'The Effingham Arms and the Empire', in *Popular Culture and Class Conflict*, eds Eileen and Stephen Yeo (Hassocks: Harvester Press, 1981), pp. 209–40; Summerfield, 'The Imperial Idea and the Music Hall', in *Imperialism and Popular Culture*, ed. John MacKenzie (Manchester: Manchester University Press, 1986), pp. 17–48; L. Senelick, 'Politics as Entertainment: Victorian Music-Hall Songs', *Victorian Studies*, 19 (1975–6), 149–80; *Music Hall: the Business of Pleasure*, ed. Peter Bailey (Milton Keynes: Open University Press, 1986), and *Music Hall: Performance and Style*, ed. J. S. Bratton (Milton Keynes: Open University Press, 1986).
18. See *Hail to the Queen: Verses for Queen Victoria's Jubilee, 1887*, comp. and intro. by Brian Louis Pearce (London: Magwood, 1987), p. 9.
19. *Ibid.*, p. 11.
20. *In the Year of the Jubilee*, ed. P. F. Kropholler (Brighton: Harvester, 1976), p. 61.
21. *Out of Work, by John Law (Margaret Harkness)*, ed. B. Kirwan (London: Merlin Press, 1990), p. 76. Further references to this edition will be shown in the text.
22. A full account of the events leading up to and consequent on 'Bloody Sunday' is to be found in E. P. Thompson, *William Morris: Romantic to Revolutionary*, rev. edn (London: Merlin Press, 1977), pp. 393–403.
23. For an account of other novels which deal with 'bloody Sunday' and the possibility of social upheaval, see my 'Conservatism and Revolution in the 1880s', in *Literature and Politics in the Nineteenth Century*, ed. John Lucas (London: Methuen, 1971), pp. 173–220.

4
'The Mote Within the Eye': Dust and Victorian Vision

Kate Flint

In 1898, the natural historian Alfred Russel Wallace published a retrospective study: *The Wonderful Century: Its Successes and its Failures*. In this, he devotes a whole chapter to 'The Importance of Dust'.[1] Dust in our towns and in our houses, he acknowledges, 'is often not only a nuisance but a serious source of disease'. As it is usually perceived by us, it is – he borrows Lord Chesterfield's original definition of dirt – 'only matter in the wrong place'. We might look to get rid of it as far as possible, for example by implementing legislation against excess or inefficient combustion of coal. But, Wallace continues:

> though we can thus minimise the dangers and the inconveniences arising from the grosser forms of dust, we cannot wholly abolish it; and it is, indeed, fortunate we cannot do so, since it has now been discovered that it is due to the presence of dust we owe much of the beauty, and perhaps even the very habitability, of the earth we live upon. Few of the fairy tales of science are more marvellous than these recent discoveries as to the varied effects and important uses of dust in the economy of nature.
>
> Wallace, *Wonderful Century*, pp. 68–9

Dust was a paradoxical substance, its position within Victorian culture perennially unstable. Emotively – and logically – it was associated with disease; its elimination or control with necessary practices of hygiene. As such, its properties are co-terminous with the wider category of dirt, and – to quote Mary Douglas's *Purity and Danger* – 'dirt avoidance for us is a matter of hygiene or aesthetics and is not related to our religion'.[2]

Yet once one accepts that not all dust is dirt, its resonances broaden out. These resonances may still remain pejorative. Thus dust may be seen as the marker of undesirable class status. Pip, in Dickens's *Great Expectations* (1861), wants to escape from Joe's forge, 'dusty with the dust of small coal'.[3] But dust is also an equalizer, as well as a factor in establishing hierarchies. Its long-standing equation with the most reductive form of matter to which we must all return – 'dust to dust' – ensures that its evocation was full of metaphorical opportunities. And its indispensable value was also perceived, both in the basic sense of waste reclamation, typified by the material value of the Harmon Mounds in Dickens's *Our Mutual Friend* (1864–5), with their 'golden dust', and in the functions that it was seen to perform within nature. Here, again, the potential for moral elaboration is unmistakable, even in the introduction to a scientific article. J. G. McPherson writes in *Longman's Magazine* in May 1891: 'Some of the most enchanting phenomena in nature are dependent for their very existence upon singularly unimportant things; and some phenomena that in one form or another daily attract our attention are produced by startlingly overlooked material'.[4] He cites the glow of an autumnal evening, the colour of the Mediterranean, the deep blue of the summer sky, 'when the eye tries to reach the absolute', mist, snow, rain, and hail. On the other hand, he asks:

> What is the source of much of the wound putrefaction, and the generation and spread of sickness and disease? What, in fact, is one of the most marvellous agents in producing beauty for the eye's gratification, refreshment to the arid soil, sickness and death to the frame of man and beast? That agent is *dust*.
>
> McPherson, 'Dust', p. 50

The paradox does not end here, in the juxtaposition of beauty and utility with disease and decay. It is a paradox crucially interwoven with the Victorian interest in the visible and the unseen. For dust gives rise to atmospheric effects which, as McPherson puts it, 'have a most important influence upon the imagination [....] [A]n aesthetic eye is charmed with their gorgeous transformation effects', since they stretch the mind towards contemplation of the vastness of space, of infinity' (McPherson, 'Dust', p. 50). But one cannot extrapolate simplistically from this and say that if one can see dust, it is to be equated with waste, with excess, with residue, yet if one cannot see it, only the effects to which it gives rise, we can appreciate its value and beauty.

Danger, as well as the potential for beauty, may well lie concealed from the human eye: individual dust particles are so tiny that 'a microscope magnifying 1,600 diameters is required to discern them', yet, McPherson writes, 'some are more real emissaries of evil than poet or painter ever conceived' (McPherson, 'Dust', p. 50). To think about dust, in other words, is to think not just about aspects of the materiality of Victorian life, but to consider debates concerning the perception of the material world and the conditions of vision that make this perception possible. Dust, both pervasive and evanescent, functions not only as a powerful literary metaphor; its specks also provide a meeting point for the intersection of science, vision and imagination.

What is dust? Nineteenth-century scientists developed a series of increasingly refined experiments to determine its composition: methods refined by Louis Pasteur, who used gun-cotton or asbestos as a filter and then dissolved it in ether, and this proved the most effective method. John Tyndall, in the late 1860s, conducted a series of experiments by means of this technique which proved, to his surprise, that a considerable proportion of the particles floating in the air of London were of organic, rather than inorganic origin. Victorian city streets were full of dust, and to remark on it was to underscore the unhealthy hostility of the urban environment. It swirls around the spring-time streets of London – 'such a gritty city; such a hopeless city' – at the opening of Chapter 12 of *Our Mutual Friend*, a novel notoriously permeated with dust imagery, and for which Dickens considered 'Dust' a possible title;[5] it characterizes the bleakness of a sandwich-board man's existence as he treads through W. E. Henley's 'Trafalgar Square' at the end of the century in 'An ill March noon; the flagstones gray with dust;/An all-round east wind volleying straws and grit'.[6] Henry Mayhew, in *London Labour and the London Poor* (1861), remarks that 'In some parts of the suburbs on windy days London is a perfect dust-mill', and records the water-carts that used to go out to damp down the streets.[7] The inorganic dust particles came from the pulverized dried mud of the streets, the wearing down of granite pavements and roadways by feet and by iron-shod horses, and above all from smoke, helping form, on occasion, what Esther Summerson, in *Bleak House* (1852–3) mistook as 'dense brown smoke' from a 'great fire'.[8] By the end of the century, it was claimed that 'No less than 350 tons of the products of the combustion of sulphur from the coal are thrown into the atmosphere of London every winter day' (McPherson, 'Dust', p. 58). The problem was even worse in some industrial towns, prompting

Ruskin's apocalyptic recognition of the 'storm-cloud of the nineteenth century', Manchester's 'sulphurous chimney-pot vomit of blackguardly cloud' spewing out a pall of pollution.[9] In addition to the inorganic materials were

> particles of every description of decaying animal and vegetable matter. The droppings of horses and other animals, the entrails of fish, the outer leaves of cabbages, the bodies of dead cats, and the miscellaneous contents of dust-bins generally, all contribute their quota to the savoury compound.[10]

But this was only half of the story. For dust was not confined to the outside. Invasively, it quickly built up in the home, a fact given a threatening spin in a curious book by H. P. Malet, *Incidents in the Biography of Dust*, where the dust particles themselves threateningly address the reader: 'At this present moment we see ourselves on the table, the books, and the inkstand; if we were not carefully removed daily, we should soon bury them, as we buried Tyre and Sidon.'[11] 'Few people have any conception of the amount of dirt contained in an ordinary carpet', the physician Robert Brudenell Carter ominously announced in 1884; 'Curtains are even worse' (Carter, 'Lighting', p. 398). Stir up a sitting-room carpet with a broom, he suggests; let the dust settle for half an hour, put it under the microscope, and what does one find? Mrs Beeton's comment that 'Nothing annoys a particular mistress so much as to find, when she comes downstairs, different articles of furniture looking as if they had never been dusted' had increasingly more than a fastidious sniff behind it as the understanding of bacterial transmission grew.[12] Even if the only bacterium to be identified with certainty by the 1890s was that which caused suppuration in wounds, hypothetical speculation from the mid-century onwards populated the air with tiny disease-bearing organisms like dangerous insect swarms. Certainly it became recognized that dust settling on food caused the multiplication of bacteria. Moreover, the body created its own dust, through the constant shedding of 'the scales of the epidermis'.[13]

Other dusty dangers lurked within the home. Florence Nightingale, in her *Notes on Nursing*, warns that certain green wallpapers give off arsenic dust:[14] it follows that those who worked in the manufacture of such papers, and in other dust-producing industries and trades suffered badly. The lungs of 'coal miners and miners in general, knife-grinders, needle-pointers, quarrymen, stonecutters, millers' – to borrow the list of a physician in the mid-1860s – were all subject to injury from dust.[15]

If Mrs Thornton, in Elizabeth Gaskell's *North and South* (1854–5), covers up her furniture against damage from the dirty Manchester air, and gazes with disapproval at the Hales' small drawing-room ('The room altogether was full of knick-knacks, which must take a long time to dust; and time to people of limited income was money'),[16] this pragmatic middle-class angst is put into perspective by Bessy Higgins telling of the conditions in the mill, where the air is full of bits of fluff, 'as fly off fro' the cotton, when they're carding it, and fill the air till it looks all fine white dust. They say it winds round the lungs, and tightens them up. Anyhow, there's many a one as works in a carding-room, that falls into a waste, coughing and spitting blood, because they're just poisoned by the fluff' (Gaskell, *North and South*, p. 146).

If dust was a hazard of the industrial city, so did it form an unfavourable aspect of colonial life. This is brought out well in Emily Eden's *Up the Country* (1866), a collection of letters written in 1837–40. The dust in India is 'much worse' than a London fog back home.[17] In Cawnpore, 'people lose their way on the plains, and everything is full of dust – books, dinner, clothes, everything' (Eden, *Up the Country*, p. 64). It is as though the substance of the country is performing a kind of reverse colonization. Anxiety about the insinuating, corrupting qualities of dust was certainly merited in terms of practical hygiene. Flora Annie Steele and Grace Gardiner, in *The Complete Indian Housekeeper and Cook*, warn that 'Dirt, illimitable, inconceivable dirt must be expected, until a generation of mistresses has rooted out the habits of immemorial years'.[18] They advise that a much larger quantity of dusters will be necessary than in an English household, and recommend tan stockings and shoes, 'as they do not hold the dust' (Steele and Gardiner, *Housekeeper*, pp. 53, 174). But this anxiety concerning hygiene is also a metaphorical lens, as Gail Low has suggested, through which a central problem of Empire – the fear of contamination of national identity – becomes evident.[19]

Mary Douglas suggests that:

> If we can abstract pathogenicity and hygiene from our notion of dirt, we are left with the old definition of dirt as matter out of place. This is a very suggestive approach. It implies two conditions: a set of ordered relations and a contravention of that order. Dirt then, is never a unique, isolated event. Where there is dirt there is system. Dirt is the by-product of a systemic ordering and classification of matter, in so far as ordering involves rejecting inappropriate elements.
>
> Douglas, *Purity and Danger*, p. 35

Once one moves away from a definition of dust as dirt, its status becomes less stable. Its position as marginal, surplus, unwanted matter may be reversed. 'Rubbish', as has been succinctly stated by Michael Thompson in his book on *Rubbish Theory*, 'is socially defined',[20] and hence one person's discarded waste can be another person's source of wealth. The arguments surrounding precisely what *was* in those mounds which dominate the plot and landscape of *Our Mutual Friend* have been well enough rehearsed, and it has been satisfactorily established that even if these heaps probably did not, after the sanitary measures of 1848, stand in proximity to where human excrement was deposited, the two were likely to have been closely associated in the popular mind.[21] Hence the symbolic relationship of wealth to shit ('dust' had been a colloquial word for money since the early sixteenth century) has been a plausible enough critical extrapolation, and, moreover, one which has received psychoanalytic endorsement through Freud's equation of money with faeces.[22] Readers of *Household Words* would already have been familiar with the idea of the value of dust – taking the word in its broadest sense. John Capper's article 'Important Rubbish' in *Household Words* classifies the contents of the mounds, thus bringing system to them, rescuing them from the category of dirt,[23] and a visit to the dust-yards, showing a concomitant fascination with recycling, became something of a mid-nineteenth-century journalistic standby. The most valuable of all components – excepting the occasional coins or pieces of jewellery – were coals, coal-dust and half-burned ashes: the 'breeze' that was baked into building blocks, and 'thus', the readers of the *Leisure Hour* were reminded in 1868, 'our houses may be said to arise again from the refuse they have cast out'.[24] Bones went to boiling-houses, to be turned into soap and gelatine, and eventually toothpicks and knife-handles and toothpowder, and fertilizer. Paper becomes papier-mâché or paper again; clothes are sent off to make shoddy – torn-up woollen material – and in turn become clothes again. This shoddy was known as 'devils-dust', ostensibly from the name of the machine used to tear up the fabrics, but in fact redolent of the poisonous nature of the greasy, germ-ridden, 'choking clouds of dry pungent dirt and floating fibres', as Mayhew termed them (Mayhew, *London Labour*, p. 30): 'Devilsdust', the double-edged name, signifying both exploitation and sedition, given by Disraeli to a dark, melancholy, ambitious, discontented ponderer on the rights of labour in *Sybil* (1845), who had started his working life as a nameless orphan manufacturing shoddy. Glass, old shoes, metals: all were re-used. And broken toys and chipped china frequently were appropriated by the women who carried out most of the dust-sifting for their own homes. A philanthropic visitor of the 1880s remarks how one home she knows 'is beautified throughout with dust-bin trophies, the mantelpiece and side-table shining with showy bits of glass and china and ornaments of various devices. There are cut-glass decanters, flower-vases, wine-glasses, tumblers, and even a delicate little bowl of the lately fashionable iridescent glass'.[25] If dust at its most pernicious is insidious, invisible, here we have dust brought into view and celebrated, commodified.

Andrew H. Miller, in *Novels behind Glass*, has usefully noted that 'Dickens' final validation of the dust-heap [...] presents the possibility that revolutionary change is unnecessary: if the potential of what we discard is actually used, then a fundamental restructuring of the economy will not be required'.[26] The metaphoric potential of dust, the extraction of value from the abject, or the restitution of the discarded, was a common trope. At its simplest, excavating a dust-heap for what is lost provides a return to order. In Wilkie Collins's *The Law and the Lady* (1875), the heroine, Valeria Woodville, gains proof that her husband did not in fact murder his former wife when she has a dust-heap excavated in the grounds of the house where they had been staying at the time of her death. This archaeological excavation in miniature digs through layers of ashes and other household refuse in order to turn up morsels of paper, which, when painstakingly restored into the form of the letter they had once been, prove Eustace Woodville's innocence, reinstate his good name, save his second marriage, and bring the complex plot to a resolution. At a more obviously figurative level, Ruskin, in *The Ethics of the Dust* (1866), provides a paradigm which was borrowed by others, and which itself was recycling paragraphs already published in *Modern Painters*, V (1860). He invites one to consider 'the dust we tread on', taking, by way of example, 'an ounce or two of the blackest slime of a beaten footpath on a rainy day, near a large manufacturing town'. Here, all kinds of geological elements are at helpless war with one another. But suppose one could separate out and in some wonderful way extract and recombine their atoms, we obtain a clear blue sapphire from the clay, an opal from the sand, a diamond from the soot, and a star-shaped drop of dew from the water. This instantly becomes a lesson in politics: 'political economy of competition' is replaced by 'political economy of co-operation'.[27] The distillation of even the most unpromising, basic raw material produces naturally formed beauty combined with material wealth. Ruskin's lesson is secular: a similar but theological point is made in Charles Reed's *Diamonds in the Dust* (1866), in which he asks his Sunday School readers to consider, among other things, 'the boyhood of great men, men who have come up from the ranks of poverty': a Smilesian list of men like Isaac Newton, Humphry Davy, James Watt, Brunel and Martin Luther: 'Are not all these from the dust of the earth, and are they not diamonds of the first water?'[28]

Other religious appropriations of dust take the reader straight back to the dust-heap. Mabel Mackintosh's children's story *Dust, Ho! or, Rescued from a Rubbish Heap* tells of a couple of girls sorting the rubbish that their drunken father collects.[29] One day, Janet comes upon a coloured picture of a seated man who had drawn a little boy close to him: beyond him a group of mothers with children, beyond them a crowd of angry men, and beneath the illustration the text 'Suffer little children to come unto Me', and a hymn. She takes the picture inside as a present to amuse her little crippled brother; a middle-class charitable visitor, who had noticed the beautiful girls in their squalid surroundings, visits their home, explains the message of the pamphlet. The words provide sustenance and hope for the dying cripple; the visiting woman gives Janet the opportunity to become a servant in her own home, and the other girl proves the means of salvation through which her father is weaned off gin and onto the gospel. Providential illumination here is literally found in the dust, where it might least be looked for; similarly, there is a message about human good being redeemable from the least promising surroundings. All these homiletic lessons derive from asking, at least implicitly, the question posed by Eustace R. Conder: 'What can seem of less consequence, or more worthless, than a pinch of dust?'[30] The reader is reminded that every one of our actions is watched; that throughout our life, we constantly leave deposits and pick things up: 'Pray that when your life-journey comes to an end, the *dust under your feet* may show that you have been walking in the right road' (Conder, *Dust*, p. 3). God here is turned into a detective-like figure, just as Sherlock Holmes reads signs of past movements in the dust in *A Study in Scarlet* (1887).

To clear away dust is to bring the past to light. But to fail to rise from the dust is to be consigned to the dust-heap of history. Ouida makes this point strongly, if a little mawkishly, in her short story 'Street Dust', in which two orphan children from the Campagna come into Rome after their mother's death to sell flowers, are arrested for begging, victimized by what Ouida portrays as a corrupt, compassionless bureaucracy, turned helpless and penniless onto the street, and take shelter in a church portico in a half-demolished street, down which a keen wind blows 'clouds of grey dust'.³¹ However, in this late-Victorian tale, there is no divine rescuing hand helping these social victims. They are found by scavengers the next morning, dead; taken to the mortuary:

> and thence, none recognizing them, they were carried to the common ditch in which the poor and nameless lie. What were they more than the dust of the street, blown about a little while by the winds, and then swept away and forgotten?
> Ouida, 'Street Dust', p. 56

It is this fear, that dust equals oblivion, a return to origins which we cannot transcend, that Tennyson seeks to redress in his poetry, most notably in 'In Memoriam' (1850). His dread is that all we will come to is 'Two handfuls of white dust, shut in an urn of brass!' as he puts it in 'The Lotos-Eaters' (1832; l. 113).³² In the early times after Hallam's death, Tennyson wants to have trust in a notion of immortality, 'Else earth is darkness at the core,/And dust and ashes all that is' (xxxiv. 3–4), but is vulnerable to the suggestion that it might be possible for a voice from beyond the grave to murmur – from some presumably earth-bound afterlife – '"The cheeks drop in; the body bows;/Man dies: nor is there hope in dust"' (xxxv. 3–4). The dread is that human existence, and the memory of it, will be subject to the same process of erosion as hills slowly eaten away by streams to create 'The dust of continents to be' (xxv. 12); subject to 'Time, a maniac scattering dust' (l. 7); that those who have loved, and suffered, and 'battled for the True, the Just', will at the end of it all 'Be blown about the desert dust' (lvi. 19), reduced to the elemental fragments that go to make up our physical composition, and deprived of all sense of identity. But in the Prologue, Tennyson has already set up God's role in all of this: 'Thou wilt not leave us in the dust' (Prologue, l. 9), he confidently announces, and goes on to piece together the atomized being.

Tennyson's fear, I suggest, is not just of the certainty and finality of mortality. It chimes with a wider fear of reversion, of history not moving in a confident forward direction, of contemporary society decaying, or alternatively sinking as a result of catastrophe or degeneration. The apprehension that haunts Tennyson, and others, is of what might happen both to an individual, and to morally bankrupt society, as apotheosized in the final line of Hopkins's 'The Sea and the Skylark' (c.1877), that we are breaking 'down/To man's last dust, drain fast towards man's first slime' (ll. 13–14).³³ This dread of reversion provides a dialectical contrast to the prevalent myth of historical progress. It is this property of dust, to remind one that the machines of the industrial age have not supplied the power to drive history forwards, that technological change is not to be equated with social betterment, that history involves the destruction as well as the accumulation of the material, that led to Walter Benjamin's fascination with dust in his vast, incomplete *Passagen-Werk* (1927–40). He quotes Henri de Pène, writing in 1859 of how he returns from the *Courses de la Marche*: '"The dust has surpassed all expectations. The elegant people back from the *Marche* are practically buried under it, just as at Pompeii; and they have to be disinterred, if not with pickaxes, then at least with a brush"'. Dust, he goes on to say, 'settles over Paris, stirs, and settles again. It drifts into the passages and collects in their corners; it catches in the velvet drapes and upholstery of bourgeois parlors; it clings to the historical wax figures in the Musée Gravin. The fashionable trains on women's dresses sweep through dust'.[34] All of this helps suggest to Benjamin that history is, at the very least, standing still: the phenomenon of dust calls the whole idea of progress, of teleology, into question.

Celebrating the beauty of history's meaningless accretions, Marcel Duchamp, as part of the visual notes that he took for the *Large Glass* (or *The Bride Stripped Bare by her Bachelors, Even*) (1915–23), fixed dust which had fallen on the surface of a flat pane of glass over a period of months: the result, photographed by Man Ray, was entitled *Elevage de poussière* (*Dust Breeding*) (1920). But it was not dust itself that was seen as possessing aesthetic potential by the Victorians, with the exception of motes dancing in the sun-beam. Even these have their demonic opposite, the motes in the moonbeam in *Dracula* (1897) which, metamorphosing into vampiric figures, show how Stoker has picked up on the poisonous, miasma-like potential of dust in the air. Rather, dust's aesthetic importance, the grounds on which dust is something to be welcomed, rested on its presumed ability to cause certain atmospheric and climatic effects.

The researches and writing of John Tyndall are crucial here. In his essay 'The Scientific Use of the Imagination' (1870), he presents his fascination with the physical basis of light to a general audience.[35] In so doing, he claims that the light of our firmament is not direct solar light, but reflected light. He elaborates on the nature of this reflection by asking why the sky is blue. He asks his audience to imagine that white solar light, as it falls, somehow gets divided, breaking into the colours of the spectrum. What he calls an 'undue fraction' of the smaller light waves are scattered by particles, particles in the air, and the proportions of this scattering ensure the predominance of the colour blue. At this point, Tyndall suggests that we consider 'sky-matter':

> Suppose a shell to surround the earth at a height above the surface which would place it beyond the grosser matter that hangs in the lower regions of the air – say at the height of the Matterhorn or Mont Blanc. Outside this shell we have the deep blue firmament. Let the atmospheric space beyond the shell be swept clean, and let the sky-matter be properly gathered up. What is its probable amount? I have sometimes thought that a lady's portmanteau would contain it all. I have thought that even a gentleman's portmanteau – possibly his snuff-box – might take it in. And whether the actual sky be capable of this amount of condensation or not, I entertain no doubt that a sky quite as vast as ours, and as good in appearance, could be formed from a quantity of matter which might be held in the hollow of the hand.
>
> Tyndall, *Imagination*, p. 36

This handful of dust excites not fear, but awe: an awe which Tyndall mediates through Kant's comment that two things fill him with this condition: 'the starry heavens and the sense of moral responsibility in man' (Tyndall, *Imagination*, p. 51). The tiny particles which go to make up 'sky-matter' are responsible for creating our sense of infinity, a sense which A.W. Moore has in turn described as partaking of the paradoxical. Infinity 'is standardly conceived as that which is boundless, endless, unlimited, unsurveyable, immeasurable', yet set against this is our own finitude. 'It is self-conscious awareness of that finitude which gives us our initial, contrastive sense of the infinite and, at the same time, makes us despair of knowing anything about it, or having any kind of grasp of it'.[36] The smallness of dust, in other words, as Tyndall uses it, has the power to create tension between our own sense of equivalent smallness, and the vastness of our physical universe: moreover, the employment of the imagination which is necessary to our comprehension of the operation of 'sky-matter' involves moving beyond our own materiality, the dust of our own bodily composition. 'Breaking contact with the hampering details of earth', Tyndall concludes his piece, the awe felt by the scientist 'associates him with a power which gives fulness and tone to his existence, but which he can neither analyse nor comprehend' (Tyndall, *Imagination*, p. 51).

Not everyone wished to accept Tyndall's theories: Ruskin, in particular, scorned them, and 'rebelled against the idea of dust-motes in the upper regions of the air, and especially resented the idea that the clear blue of the sky could be due to anything so gross and terrestrial as dust'.[37] Although their veracity was being questioned by the end of the century in respect to what *actually* causes the blue of the sky, Tyndall's arguments were rehearsed again by Wallace in *The Wonderful Century*, who explained, moreover, the aesthetic pleasure created by the effect of thicker dust particles in the lower atmosphere, particularly when struck by the slanting rays of the setting sun, producing 'not unfrequent exhibitions of nature's kaleidoscopic colour painting'. The most spectacular effects are produced when the sun has slid below the horizon, and when there are a certain quantity of clouds:

> These, as long as the sun was above the horizon, intercepted much of the light and colour; but, when the great luminary has passed away from our direct vision, his light shines more directly on the under sides of all the clouds and air strata of different densities; a new and more brilliant light flushes the western sky, and a display of gorgeous ever-changing tints occurs which are at once the delight of the beholder and the despair of the artist. And all this unsurpassable glory we owe to – dust!
>
> Wallace, *Wonderful Century*, pp. 73–4

These theories had been confirmed, for Wallace, by the explosion of Krakatoa on 26–7 August 1883 which had released, it was estimated, some 70,000 cubic yards of dust into the atmosphere. These circled the globe several times over the succeeding years, causing the spectacular sunsets of the 1880s.[38] Gerard Manley Hopkins, in one of his rare appearances in print, contributed to a correspondence in the science journal *Nature* recording these recent phenomena, before their cause was ascertained.[39] He notes how they differed from ordinary sunsets, the light being both more intense and yet lacking in lustre, the colours being impure and not of the spectrum. His account of the sunset of 16 December 1883 demonstrates his Ruskin-influenced techniques of precise observation as he notes:

> A bright glow had been round the sun all day and became more remarkable towards sunset. It then had a silvery or steely look, with soft radiating streamers and little colour; its shape was mainly elliptical, the slightly longer axis being vertical; the size about 20 from the sun each way. There was a pale golden colour, brightening and

> fading by turns for ten minutes as the sun went down. After the sunset the horizon was, by 1.10, lined a long way by a glowing tawny light, not very pure in colour and distinctly textured in hummocks, bodies like shoals of dolphins, or in what are called gadroons, or as the Japanese conventionally represent waves.
>
> Hopkins, Correspondence, p. 165

So strange are these solar manifestations that he is resorting to similes drawn not from other aspects of nature, but from stylized representation. Such sunsets fed directly into poetry: into, for example, Tennyson's 'St Telemachus' (1892) and *Eros and Psyche* (1885), by Hopkins's friend Robert Bridges, a piece which has some suspiciously close verbal resemblances to Hopkins's own published account.

Despite the emphasis which Wallace placed on the aesthetic appeal of dust's effects, he was ready to concede that there might be some who would be willing to sacrifice them if by doing this they would be escaping its disagreeable properties. But dust is not dispensable. He, like other late-nineteenth-century commentators on the topic, calls attention to the work of the Scottish scientist John Aitken, who proved that it is the presence of dust in the higher atmosphere that causes 'the formation of mists, clouds, and gentle beneficial rains, instead of waterspouts and destructive torrents' (Wallace, *Wonderful Century*, p. 76). This, together with its capacity to produce beauty, allows Wallace to make a strong case for the rehabilitation of dust's reputation. Despite the fact that it brings dirt, discomfort, and even disease, it is 'an essential part of the economy of nature', both helping to render life more enjoyable in aesthetic terms, and being nothing less than essential to our climatic systems. From this, he draws a conclusion very similar to other, more pious commentators: 'The overwhelming importance of the small things, and even of the despised things of our world, has never, perhaps, been so strikingly brought home to us as in these recent investigations into the wide spread and far-reaching beneficial influences of Atmospheric Dust' (Wallace, *Wonderful Century*, p. 83).

To focus on dust, as I suggested earlier, raises certain questions about Victorian fascination with the relationship between the visible and the invisible, and with techniques of seeing, both technological and physiological. Added to this must be concern with the individuality which manifests itself in the act of seeing and the recording of this act: what

G. H. Lewes wrote of as our subjective co-operation in the perception of objects, or, to recast this through George Eliot's words in *Middlemarch* (1871–2): 'Will not a tiny speck very close to our vision blot out the glory of the world, and leave only a margin by which we see the blot? I know no speck so troublesome as self'.[40] This analogy, employed to show the impossibility of stable, objective vision, draws on the sense that dust is simultaneously indispensable yet problematic.

The study of dust and its effects, both injurious and beneficial, would have been impossible without the developments which took place in the technology of the microscope, unlocking, as Philip Gosse put it, 'a world of wonder and beauty before invisible'.[41] Once again, the popular literature which developed around the domesticized version of this instrument tended to emphasize the divinely sanctioned social messages to be gleaned from concentrating on the miniature and the obscure. Mary Ward, for example, in *A World of Wonders Revealed by the Microscope* reminds the 'Emily' to whom this work is ostensibly addressed that one is looking at 'the works of One who judges not as we do of great and small; who "taketh up the isles as a very little thing," and counts the nations as "the small dust of the balance"', yet promises individual salvation to each being from those nations.[42] More generally, the microscope was praised for its ability to train one's powers of careful observation, and for its democracy – a considerable amount of useful work could be performed by the amateur naturalist, it was asserted, with a very cheap instrument; this work could be performed by those living the most cramped of urban existences, taking the raw materials of their science from the world around them, investigating 'the commonest weed or the most familiar insect […]. There is not a mote that dances in the sunbeam, not a particle of dust that we tread heedlessly below our feet, that does not contain within its form mines of knowledge as yet unworked. For if we could only read them rightly, all the records of the animated past are written in the rocks and dust of the present'.[43] We may come to see through '"the world of small"', to use William Carpenter's term, that size is relative, that mass has nothing to do with real grandeur. 'There is something', he continues, 'in the extreme of minuteness, which is no less wonderful, – might it not almost be said, no less majestic? – than the extreme of vastness'.[44]

Yet whether examining the minuscule, or the vastness of the heavens, the more one could see when the natural powers of the eye were augmented by the crafted lens, the more scientists were aware of what lay beyond one's visual reach. It was here that observation of the natural world had to yield place to the importance of the imagination, increasingly recognized as having a central role within scientific inquiry, taking one beyond what the eye can see. The imagination is the instrument with the true power to open things up. As G. H. Lewes, one of the most sustained advocates for the employment of this faculty, put it: 'The grandest discoveries, and the grandest applications to practice, have not only outstripped the slow march of Observation, but have revealed by the telescope of Imagination what the microscope of Observation could never have seen.'[45] Tyndall invited his reader to:

> Conceive a grain of sand of such a size as just to cover the dot placed over the letter *i* in these pages; there are animals so small that whole millions of them, grouped together, would not be equal in size to such a grain of sand. These are the results of microscopic research; but the microscope merely opens the door to imagination, and leaves us to conjecture forms and sizes which it cannot reveal.[46]

What we are being encouraged to do, in other words, is to learn to see differently, to see with the mind's eye. The powerful lens of the microscope, revealing simultaneously the dangers and the welcome properties of dust, is not enough in itself. Nor is it sufficient to see with the eye of the social recorder, although this may allow one to bring order to the components of dust and hence, in Mary Douglas's terms, reclaim the properties of the dustheap from over the borderline of that which has been discarded, and which hence threatens social order. Dust, as so many commentators on the materiality of this substance pointed out, is a paradoxical substance: a threat, yet, to use a formulation of Wallace once again, 'a source of beauty and essential to life' (Wallace, *Wonderful Century*, p. 68). But its real fascination to the Victorians lay not so much in the dialectics of this materiality, but in the fact that its insidious physical presence also partook of something far more metaphysical; reached, even, towards the Kantian sublime. As Tyndall acknowledged in 1870, 'beyond the present outposts of microscopic enquiry lies an immense field for the exercise of the speculative power' (Tyndall, *Imagination*, p. 41). The importance of dust to Victorian culture lies precisely in this capacity to suggest the vastness of imaginative conjecture that may lie behind and beyond the most apparently mundane: the invisible behind the visible.

Notes

1. (London: Swan Sonnenschein, 1898), pp. 68–84.
2. *Purity and Danger: an Analysis of the Concepts of Pollution and Taboo* (London: Routledge, 1966; 1984), p. 35.
3. (Harmondsworth: Penguin, 1965), p. 135.
4. J. G. McPherson, 'Dust', *Longman's Magazine*, 18 (1891), 49–59 (p. 49).
5. Charles Dickens, *Our Mutual Friend*, ed. Stephen Gill (Harmondsworth: Penguin, 1971), p. 191.
6. W. E. Henley, 'Trafalgar Square', in William Nicholson, *London Types: Quatorzains by W. E. Henley* (London: Heinemann, 1898), p. [vii]. The sonnet illustrates Nicholson's print of a sandwich-man advertising Seeley's *Ecce Homo*.
7. Henry Mayhew, *London Labour and the London Poor*, 4 vols (London: Griffin, 1861), II, 188.
8. (New York: Norton, 1977), p. 29. For an overview of this topic, see Peter Brimblecombe, *The Big Smoke: a History of Air Pollution in London since Medieval Times* (London: Methuen, 1987).
9. John Ruskin, *The Storm-Cloud of the Nineteenth Century*, Lecture I (1884), in *Complete Works*, eds E. T. Cook and A. Wedderburn, 39 vols (London: George Allen, 1903–12), XXXIV, 38.
10. Robert Brudenell Carter, 'Lighting', in *Our Homes, and How to Make them Healthy* (London: Cassell, Petter, Galpin, 1883–5), pp. 397–8.
11. (London: Trübner, 1877), p. 34.
12. Mrs Isabella Beeton, *The Book of Household Management* (London: S. O. Beeton, 1861), p. 1001.
13. [Anon.], 'Dust and Hygiene', *All the Year Round*, 3rd series, 13 (1895), 154.
14. *Notes on Nursing for the Labouring Classes* (London: Harrison, 1868), p. 73.
15. F. Oppert, *On Melanosis of the Lungs and Other Lung Diseases arising from the Inhalation of Dust* (London: John Churchill, 1866), pp. 3–4.
16. (Harmondsworth: Penguin, 1970), p. 139.
17. Emily Eden, *Up the Country* (London; Virago, 1983), p. 110.
18. Two Twenty Years' Residents [Flora Annie Steele and Grace Gardiner], *The Complete Indian Housekeeper and Cook* (Edinburgh: Frank Murray, 1890), p. 57.
19. Gail Low, *White Skins, Black Masks* (London: Routledge, 1996), p. 162.
20. *Rubbish Theory: the Creation and Destruction of Value* (Oxford: Oxford University Press, 1979), p. 11.
21. See Harvey Peter Sucksmith, 'The Dust-heaps in *Our Mutual Friend*', *Essays in Criticism*, 23 (1973), 206–12.
22. See Sigmund Freud, 'Character and Anal Eroticism' (1908), *The Standard Edition of the Complete Psychological Works of Sigmund Freud*, ed. James Strachey and others, 24 vols (London: Hogarth Press and the Institute of Psychoanalysis,1953–74), IX (1959), 169, and 'The Interpretations of Dreams', *Works*, V (1953), 403.
23. 11 (1855), 376–9.
24. [Anon], 'Our Dust-bins', *Leisure Hour*, 17 (1868), 719.
25. H. A. Forde and her sisters, *Dust, Ho! and Other Pictures from Troubled Lives* (London: Christian Knowledge Society, *c.* 1885), p. 13.

26. *Novels Behind Glass: Commodity Culture and Victorian Narrative* (Cambridge: Cambridge University Press, 1995), p. 125.
27. John Ruskin, *Modern Painters*, V (1860), in *Works*, VII, 207.
28. *Diamonds in the Dust: a New Year's Address for Sunday Scholars* (London: Sunday School Union, 1866), pp. 21–2.
29. (London: John F. Shaw, 1891).
30. Eustace R. Conder, *Dust; and Other Short Talks with Children* (Leeds: W. Brierly; London: Hodder and Stoughton, 1882), p. 1.
31. *Street Dust and Other Stories* (London: F.V. White, 1901), p. 50.
32. All quotations from Tennyson's poetry are taken from *The Poems of Tennyson*, ed. Christopher Ricks, 3 vols (London: Longman, 1987).
33. Quotations from Hopkins's poetry are taken from *The Poetical Works of Gerard Manley Hopkins*, ed. Norman H. MacKenzie (Oxford: Clarendon Press, 1990).
34. Susan Buck-Morss, *The Dialectics of Seeing: Walter Benjamin and the Arcades Project* (Cambridge, MA: MIT Press, 1989), pp. 95–6.
35. John Tyndall, *Essays on the Use and Limit of the Imagination in Science* (London: Longman, Green, 1870).
36. *Infinity*, ed. A.W. Moore (Aldershot: Dartmouth Publishing, 1993), p. xi.
37. Oliver Lodge, 'Ruskin's Attitude to Science', *St George*, 8 (1905), 290; quoted in Ruskin, *Works*, XXXVII, 525.
38. See further Richard D. Altick, 'Four Victorian Poets and an Exploding Island', *Victorian Studies*, 3 (1960), 249–60; Thomas A. Zaniello, 'The Spectacular English Sunsets of the 1880s', in *Victorian Science and Victorian Values: Literary Perspectives*, eds Jim Paradis and Tom Postelwait, *Annals of the New York Academy of Sciences*, 360 (1981), 247–67.
39. Gerard Manley Hopkins, 'The remarkable sunsets', *Nature*, 29 (3 January 1884), 222–3; repr. in *The Correspondence of Gerard Manley Hopkins and Richard Watson Dixon*, ed. Claude Colleer Abbott (London: Oxford University Press, 1935), pp. 161–6.
40. (New York: Norton, 1977), p. 289.
41. Philip Henry Gosse, *Evenings at the Microscope; or, Researches Among the Minuter Organs and Forms of Animal Life* (London: Society for Promoting Christian Knowledge, 1859), p. iii.
42. The Hon. Mrs W. [Mary Ward], *A World of Wonders Revealed by the Microscope* (London: Groombridge, 1858), p. 8.
43. Rev. J. G. Wood, *Common Objects of the Microscope* (London: Routledge, Warne, and Routledge, 1861), p. 3.
44. William B. Carpenter, *The Microscope: and its Revelations* (London: John Churchill, 1856), p. 36.
45. G. H. Lewes, *Problems of Life and Mind*, Series I (London: Trübner, 1874), p. 315.
46. John Tyndall, *Natural Philosophy in Easy Lessons* (London: Cassell, Petter and Galpin, 1869), p. 6.

5
Purging Christianity of its Semitic Origins: Kingsley, Arnold and the Bible

Stephen Prickett

Few members of the diplomatic corps of a foreign country can have played a larger part in the cultural life of the host nation than Baron Christian Bunsen, the Prussian ambassador to the Court of St James from 1841 to 1854. In 1847 he was to add a personal alliance to his political mission to Britain by his marriage to Frances Waddington, a Welsh heiress, but his place in British life had started much earlier and extended far beyond the diplomatic and social sphere. In 1841 he inaugurated his long-cherished scheme for union between the Anglican and Prussian Lutheran State church with a proposal for a joint Bishopric of Jerusalem. Each Church would appoint its candidate in turn, and the Bishop then chosen would minister alike to both Anglican and Lutheran communities. The idea was taken up enthusiastically not merely among the liberal Anglicans, who were closest theologically to Bunsen, but even initially by Pusey and the Tractarians – and was effected by an Act of Parliament in October of that year.[1] Geoffrey Faber has argued that, had Newman not in the end been so implacably opposed to the whole scheme, helping to make it a one-off arrangement rather than the prelude to what some hoped would be eventual union, the relations between Britain and Prussia might have been so different as to avert the catastrophe of the First World War.[2] Certainly Bunsen was a key figure in London intellectual life of the 1840s, his ideas tapping into a huge existing reservoir of pro-Teutonic sentiment.

Though it is tempting (if not quite accurate) to suggest that English 'Teutomania' flourished in inverse proportion to actual knowledge of the German language, it would certainly be true to note that enthusiasm for Teutonic culture was out of all proportion to a detailed acquaintance with the country and its inhabitants. While Germany was frequently contrasted with France, which was vilified as militarily dangerous, aggressively Catholic, sexually *risqué* and signally lacking in serious or profound thinkers, it is nevertheless remarkable how often Germany was actually seen through French eyes. Even Arnold himself often seems to have gathered his awareness of German thought more from the pages of Renan, or the *Revue des deux Mondes*, than from a detailed reading of Hegel or Schleiermacher. Closer inspection, moreover, was sometimes less starry-eyed. Thomas Hood's *Up the Rhine* (1839) had managed to suggest that though the Germans were lazy, dirty and dishonest, the countryside was beautiful and the cost of living – for anyone fortunate enough to pay in pounds – extraordinarily low. Victorian novelists echo the refrain that Germany, though unfashionable, is at least cheap – as witnessed by the number of disgraced fictional protagonists who depart to eke out their days there in exile. Such snide on-the-ground observations, however, were easily outweighed by the literary, philosophical and musical prestige of Goethe and Schiller, Kant and Hegel, Beethoven, Brahms, Schubert and Mendelssohn. Admiration for Germany as the home of profound (if sometimes incomprehensible) genius was reinforced by the tireless propaganda of Carlyle, Kingsley and (from America) Emerson, while in the mid-years of the century London society, already marked by the evident seriousness of the Prince Consort, was even more impressed by Bunsen's advocacy of a natural theological kinship between the English and the land of Luther and Schleiermacher.

Not least of the attractions of this appeal to an innate ethnic and spiritual bond between the two peoples was its implications for one of the most delicate theological problems of the day: what to do with the Bible. Victorian attitudes towards the Bible were nothing if not ambiguous. If, for millions, it was still the Inspired Word of God, speaking directly to the hearts and minds of its faithful (and, it was assumed, mostly Protestant) readers, more and more of those millions were becoming uncomfortably aware of the lengthy processes of mediation by which it had reached them. Robert Lowth's translation of Isaiah, one of the few new translations of the eighteenth century, had stressed that its primary goal was to provide a sounder text for personal and figurative interpretations, but even to call attention to the fact that the wording of the English Authorized Version was neither inspired nor immutable, raised awkward wider questions about the origins and reliability of the sacred text itself.[3]

Though there is a sense in which the historical criticism of the Bible had begun in England in the early eighteenth century, with a few exceptions, it had failed to take root there as it did in Germany.[4] In the early nineteenth century the intellectual isolation created by the Revolutionary and Napoleonic Wars was compounded by a truly massive absence of German teaching in schools and universities. Carlyle's German, like Coleridge's, was self-taught. In the Divinity Schools German was not studied: when, in 1823, Pusey, the future Regius Professor of Hebrew, wanted to learn about developments in Lutheran scholarship, he discovered that there were only two men in the University of Oxford capable of reading German.[5] Cambridge was marginally better off, but even in Scotland, Glasgow, for instance, did not appoint its first lecturer in German until 1858 – only three years before it bowed to the external pressure of a Royal Commission and created its first Chair of English Literature.[6] The impact of George Eliot's translations of Strauss's *Life of Jesus* (1846) and Feuerbach's *Essence of Christianity* (1852) was, therefore, the more disturbing.

Yet it is revealing that when, as late as 1862, the liberal-minded Bishop of Natal, John Colenso, published the first volume of his *Critical Examination of the the Pentateuch* (1862–79), arguing (what is now a critical commonplace) that the first five books of the Bible were not by Moses at all, but dated from post-exilic times, and were therefore technically in this sense 'forgeries', the Bishop of Cape Town excommunicated him. The fact that he was later re-instated indicates less a transformation of the scholarly climate than the degree of nervous muddle over the origins of their faith still pervading most sections of English religious opinion. Though the oft-repeated story of Huxley confronting and routing Samuel Wilberforce, Bishop of Oxford, in debate over Darwin's *Origin of Species* (1859) at the Pitt-Rivers Museum, belongs to a tradition of mythology as potent (and as unfounded) as any related in the Old Testament, it is deeply symptomatic of an age whose traditional grand narratives were all too palpably crumbling, and which was desperately in search of new alternatives.

Yet even if, as Adrian Desmond and James Moore have both argued, Huxley and his gang of scientific myrmidons were deliberately setting out to challenge and even overthrow what they saw as the stultifying weight of the religio-political establishment, they remained a tiny minority.[7] For most readers of intellectual journals of the period – to look no wider – science provided not so much a new metanarrative as a methodology for reconstructing and rehabilitating a religion whose cultural roots penetrated so deeply into their own sense of identity and purpose as to make it almost indispensable. Like many nineteenth-century cathedrals, though the foundations were rotting away and perceptibly subsiding, the superstructure was so aesthetically and culturally attractive as to make demolition quite out of the question. What was needed, as Arnold was to argue repeatedly, was a way of dispensing with the 'aberglaube' of the old supernatural bases of Christianity and replacing them with a satisfactorily scientific foundation. Thus the distinction between 'Hebrew' and 'Hellene' in *Culture and Anarchy* (1869) is less a picturesque balancing of terms than part of a long-term programme for hellenizing, or, better still, 'indo-europeanizing' established Christianity. Victorian Church 'restoration' came in a variety of packages. For many, especially among the friends of Bunsen, the Germany of Lessing, Schleiermacher, Strauss and Feuerbach was seen as providing not so much a theological threat, as a scientific solution.

The context had been well prepared. Long before the publication of Gobineau's famous *Essay on the Inequality of Races* (1854), Carlyle, whose hero-worship was as much racial as individual, had been preaching the obvious superiority of the Teutons. These included not merely the Scandinavians, the Germans and the Dutch, but also the English (not to mention the Scots) and the Americans. They, as much as the Greeks, had had their heroic age. The *Nibelungen Lied* was the 'German Iliad' and Worms, their holy city, as venerable as Thebes or Troy. Carlyle's heroes, Eric Blood-axe, King Blue-tooth, Sven Double-Beard or Olaf the Thick were as inspiring as any warriors in Homer. But classical parallels with the Teutons were as nothing to the biblical.

The point was made forcefully and explicitly by Kingsley, whose novel *Alton Locke* (1850) contains the most extraordinary dream-sequence of the Aryan migrations westward. The same year, during a long-anticipated visit to Germany, he had seen the Roman amphitheatre and other remains at Trier. This vision of the classical world through essentially Teutonic spectacles was to find its first expression in *Hypatia* (1853). The genre of early-Christian novel was already well established by mid-century, but this is early Christianity with a difference. Kingsley's plot is taken directly from Gibbon:

> Hypatia, the daughter of Theon the mathematician, was initiated in her father's studies; her learned comments have elucidated the geometry of Apollonius and Diophantus; and she publicly taught,

both at Athens and Alexandria, the philosophy of Plato and Aristotle. In the bloom of beauty, and in the maturity of wisdom, the modest maid refused her lovers and instructed her disciples; the persons most illustrious for their rank or merit were impatient to visit the female philosopher; and Cyril beheld with a jealous eye the gorgeous train of horses and slaves who crowded the door of her academy. A rumour was spread among the Christians that the daughter of Theon was the only obstacle to the reconciliation of the praefect and the archbishop; and that obstacle was speedily removed. On a fatal day, in the holy season of Lent, Hypatia was torn from her chariot, stripped naked, dragged to the church, and inhumanly butchered by the hands of Peter the reader and a troop of savage and merciless fanatics: her flesh was scraped from her bones with oyster-shells, and her quivering limbs were delivered to the flames.[8]

Such a combination of sex, violence and religion could hardly fail, and Kingsley brings to this heady brew other ingredients all his own. The subtitle, 'New Foes with an Old Face' is enough to alert the reader to possible contemporary resonances in this bloodthirsty tale of fifth-century Alexandria, while the Preface reinforces this impression:

> I cannot hope that these pages will be altogether free from anachronisms and errors. I can only say that I have laboured honestly and industriously to discover the truth, even in its minutest details, and to sketch the age, its manners, and its literature, as I found them – altogether artificial, slipshod, effete, resembling far more the times of Louis Quinze than those of Socrates and Plato.[9]

Exactly what is 'artificial, slipshod, and effete' is made abundantly clear in the following pages: it is the Church of Rome, in its own way as poisonous as the civilization it had ostensibly supplanted. The whole action of the plot circles around an apocalyptic sense of a doomed and dying world. It is the theme of the opening paragraphs, as of the closing, and the action is punctuated by recurring references to the inevitable destruction of this whole society. The defeat of Heraclian's Roman expedition, for instance,

> was but one, and that one of the least known and most trivial, of the tragedies of that age of woe; one petty death-spasm among the unnumbered throes which were shaking to dissolution the Babylon

of the West. Her time had come. Even as St John beheld her in his vision, by agony after agony, she was rotting to her well-earned doom. Tyrannizing it luxuriously over all nations, she had sat upon the mystic beast.[10]

The Rome of Heraclian and the Vatican have been effectively elided. Though within this rotting imperial hulk Christianity offers a new life, it is emphatically not a case of salvation through the Church. 'The Egyptian and Syrian Churches', explains Kingsley, 'were destined to labour not for themselves, but for us'.[11] It is the Goths, huge, blond, child-like men, capable both of extraordinary ferocity and great kindliness, who are the future. They represent one of the strangest and (in retrospect) least attractive parts of Kingsley's theory of history, spelled out most fully in his later Cambridge lectures, *The Roman and the Teuton* (1864). For Kingsley, these pagan barbarians are the future, bringing a new vigour and vision to the effete world of the Mediterranean. Above all (except when indulging in a little local rapine, or carrying with them their own floating brothel on a Nile cruise), they have a reverence for women and a belief in monogamy that will eventually find its true expression in North European Protestantism: especially Lutheranism and the Church of England.

> Those wild tribes were bringing with them into the magic circle of the Western Church's influence the very materials which she required for the building up of a future Christendom, and which she could find as little in the Western Empire as in the Eastern; comparative purity of morals; sacred respect for women, for family life, law, equal justice, individual freedom, and, above all, for honesty in word and deed; bodies untainted by hereditary effeminacy, hearts earnest though genial, and blest with a strange willingness to learn, even from those they despised.[12]

The result is one of the most striking examples of the kind of double time-schemes of the Victorian novel noted by John Sutherland.[13] All historical novels tell us as much about the time when they were written as about the time they portray, but few are as open and explicit about it as *Hypatia*. The Catholicism of fourth-century Alexandria is, in its own way, as outmoded and doomed as the brutal paganism it replaced. It is, in effect, only a proto-Christianity; a *mélange* of eastern superstition, wily oriental politics and imperial dogmatism, saved only by the presence within it of certain forward-looking characters who dimly foreshadow, as it were, the coming age of Teutonic Protestantism a thousand years in the future. Among the historical characters with a walk-on part, only Augustine, that great pre-Romantic recorder of introspective self-consciousness, and Synesius, who only accepted the Bishopric of Cyrene on condition that he could keep his wife, are allowed to make the grade. Among the fictional protagonists, Philammon, a handsome young monk who leaves his monastery to see the world, and Raphael Ben-Ezra, a young Jew who has already seen too much of it, begin a spiritual pilgrimage towards a kind of religion as yet not invented. Both are appalled by the superstition and obscurantism they see around them, both reject the contemporary ideal of celibacy in favour of a manly, heterosexual and inner-directed morality that the reader knows is still too far in the future to be of help to them. They exemplify those historical characters who, as Mrs Humphry Ward was to observe,

> represent a forward strain, who belong as it seems to a world of their own, a world ahead of them. To them [the historian] stretches out his hand: '*You*', he says to them, 'though your priests spoke to you not of Christ, but of Zeus and Artemis, *you* are really my kindred!' But intellectually they stand alone. Around them, after them, for long ages the world 'spake as a child, felt as a child, understood as a child.'

But, as Kingsley's narrative makes clear, the Augustines or the Raphael Ben-Ezras are a tiny minority. For Mrs Ward the difference between the chaos and superstition of the past and the enlightenment of the nineteenth century can be summed up in a single word: 'Science'.

> Then he sees what it is that makes the difference, digs the gulf. '*Science*', the mind cries, '*ordered knowledge*'. And so for the first time the modern recognizes what the accumulations of his forefathers have done for him. He takes the torch which man has been so long and patiently fashioning to his hand, and turns it on the past, and at every step the sight grows stranger, and yet more moving, more pathetic. The darkness into which he penetrates does but make him grasp his own guiding light the more closely. And yet, bit by bit, it has been prepared for him by these groping half-conscious generations, and the scrutiny which began in repulsion and laughter ends in a marvelling gratitude.[14]

Mrs Ward, we recall, was Matthew Arnold's niece, and though this was written in 1888, it echoes in fictional form an attitude towards the Bible and a faith in the rational powers of science displayed in *St Paul and Protestantism* (1870), *Literature and Dogma* (1873) and *God and the Bible* (1875).

For Kingsley, as for Arnold and for Mrs Ward, Christianity does not embody a once-for-all revelation, but itself progresses and is modified by the development of the rest of human knowledge. It is the *progress* of our religious awareness that is of crucial importance – an idea whose origins were to be found in such German figures as Reimarus, Lessing, Schiller and Feuerbach.[15]

But not merely did the Germans provide a new model of developmental, evolutionary religion, but (together with the French) they also offered the first intimations of a new science that would help to shape the reformed and revitalized religion of the future: that of 'ethnography', the scientific description of races and cultures. Significantly, the OED dates the word (as of specifically German origin) from 1834. The new science struck an immediate chord in Arnold. Divisions between members of the European family, such as the Celt and the Teuton, were now as nothing compared with the great racial divide between the Indo-European and the Semitic peoples. In *Culture and Anarchy*, Arnold takes up the theme with enthusiasm:

> Science has now made visible to everybody the great and pregnant elements of difference which lie in race, and in how signal a manner they make the genius and history of an Indo-European people vary from those of a Semitic people. Hellenism is of Indo-European growth, Hebraism is of Semitic growth; and we English, a nation of Indo-European stock, seem to belong naturally to the movement of Hellenism.[16]

Two years earlier, in a letter to his mother written on Christmas Day 1867, Arnold is more explicit, suggesting his biblical writings were the continuation of a family tradition:

> I have been reading this year in connexion with the New Testament a good deal of Aristotle and Plato. Bunsen used to say that our great business was to get rid of all that was purely Semitic in Christianity, and to make it Indo-Germanic, and Schleiermacher that in the Christianity of us Western nations there was really much more of Plato and Socrates than of Joshua and David; and, on the whole,

papa worked in the direction of these ideas of Bunsen and Schleiermacher, and was perhaps the only powerful Englishman of his day who did so.[17]

Similarly, in the essay *On the Study of Celtic Literature*, echoing Renan, he explains that

> The modern spirit tends more and more to establish a sense of native diversity between our European bent and the Semitic bent, and to eliminate, even in our religion, certain elements, as purely and excessively Semitic, and therefore, in right, not combinable with our European nature, not assimilable by it.[18]

This seemingly loose reference to 'the modern spirit' is a reference to what any non-scientific nineteenth-century observer might well feel was a formidable body of scientific theory and evidence. Behind both Arnold and his mentor, Renan, at this point is the influence of two notable French authorities: Count Joseph Arthur de Gobineau and Émile Burnouf. Gobineau's four-volume *Essai sur l'inégalité des races humaines* (1853–5) had divided humanity into three main racial groups: the white, the yellow and the black. The whites, excelling in almost all physical, mental and moral qualities, alone had the qualities necessary for creating great civilizations. The yellow peoples, admittedly, had certain materialistic practicalities, but were sadly deficient in energy, and tended towards mediocrity. Lowest of all in this hierarchy were the blacks, who though possessed of no intellect, had some artistic talent. Supreme among the white races were the Aryans, but their native vigour had been sadly undermined by miscegenation with inferior races – particularly with the Semites, a mixture of white and negro stock. According to Gobineau the entire Mediterranean basin – including Italy, France, Spain, and even Portugal – had been contaminated by this unfortunate Semitic strain. Only the Aryans of northern Europe were relatively free from this taint – which naturally accounted for their superiority.

Both Arnold and Renan were heavily influenced by Gobineau's theories, not merely as to the racial hierarchies of the world, but also as to their religious fruits. Gobineau's second great work, *The Religions and Philosophies of Central Asia* (1865) is the immediate basis of Renan's *The Zeaziehs of Persia*, and of Arnold's *A Persian Passion Play*. More significantly, however, it underlies the biblical studies of both writers, culminating for Renan in the *Vie de Jesus* (1863) and for Arnold with

Literature and Dogma, God and the Bible, and *St Paul and Protestantism.* In one of his early essays, 'The History of the People of Israel' (1887–94), Renan explains that modern philology has demonstrated a kind of double movement in the history of civilizations. The Indo-Europeans, that is the people of India, Persia, the Caucasus, and Europe, were the originators of all the great intellectual, political and military achievements in world history. The Semites, which included the peoples of the near East as far as the Euphrates, however, had founded all the great religious and ethical movements. This was less a matter of reasoning, than of a 'primitive intuition' in which they excelled. Outside their religion, the Jews had neither politics, art, philosophy, or science; the very nature of their language made abstraction unknown and metaphysics impossible.[19]

This was a point to which Renan was to return in his *Life of Jesus,* where the insight of the earlier essay is extended into the central theoretical platform on which the explanation for the phenomenon of Jesus and the rise of Christianity is based:

> The poetry of the soul, faith, liberty, virtue, devotion, made their appearance in the world with the two great races which, in one sense, have made humanity – viz. the Indo-European and the Semitic races. The first religious intuitions of the Indo-European race were essentially naturalistic. But it was a profound and moral naturalism, a loving embrace of nature by man, a delicious poetry, full of the sentiment of the Infinite – the principle, in fine, of all that which the Germanic and Celtic genius, of that which a Shakespeare and a Goethe, should express in later times. It was neither theology nor moral philosophy – it was a state of melancholy, it was tenderness, it was imagination; it was, more than all, earnestness, the essential condition of morals and religion. The faith of humanity, however, could not come from thence, because these ancient forms of worships had great difficulty in detaching themselves from Polytheism, and could not attain to a very clear symbol. It was the Semitic race which has the glory of having made the religion of humanity.[20]

Here we note not merely the racial schemata of Gobineau, behind so much of Renan's and Arnold's approach to the Bible, but also even that 'Indo-European' quality of melancholy that is a repeated theme of Romantic theologians from Schleiermacher onwards and which, so notoriously, was to constitute an essential ingredient of the Arnoldian cultural syndrome. But there is another hidden agenda here that even Gobineau did not dream of. A recurring theme throughout the *Life of Jesus* is the physical contrast between the verdant beauties of lakeside Galilee and the stony harshness of Jerusalem and its desert environs. Jesus is seen as having been nurtured in a kind of Wordsworthian rural idyll, which in turn influenced his teachings.

> His preaching was gentle and pleasing, breathing nature and the perfume of the fields. He loved the flowers, and took from them his most charming lessons. The birds of heaven, the sea, the mountains, and the games of children furnished in turn the subject of his instructions.[21]

In contrast Jerusalem 'was then nearly what it is today, a city of pedantry, acrimony, disputes, hatreds, and littleness of mind. Its fanaticism was extreme, and religious seditions very frequent'.[22] This intellectual aridity was similarly reflected in the landscape. 'The parched appearance of nature in the neighbourhood of Jerusalem must have added to the dislike Jesus had of the place. The valleys are without water; the soil arid and stony.'[23] Moreover, this rigidity and fanaticism is not just a feature of Judaism, but of all unalleviated Semitic religion – demonstrated by the contemporary condition of Islam in Jerusalem.

> When Omar entered into Jerusalem, he found the site designedly polluted in hatred of the Jews. It was Islamism – that is to say, a sort of resurrection of Judaism in its exclusively Semitic form – which restored its glory. The place has always been anti-Christian.[24]

Yet another contrast which Renan pointedly makes between Galilee and Jerusalem is over priesthood. 'The priest, by his office, ever advocates public sacrifice, of which he is the appointed minister; he discourages private prayer, which has a tendency to dispense with his office.' Jesus, in contrast, 'despised all religion which was not of the heart'.[25] His was a 'pure worship, a religion without priests and external observances, resting entirely on the feelings of the heart, on the imitation of God, on the direct relation of the conscience with the heavenly Father'.[26] Whatever its historical accuracy, such a combination of traditional Gallic anti-clericalism and the emotive Protestantism of Schleiermacher would have been as immediately recognizable (if not acceptable) to Renan's audience as it was to Arnold.

In short, declares Renan, 'Jesus was no longer a Jew'.[27] If we are tempted to read this at face value, however, as simply a description of his philosophical position, we miss the thrust of the geographical and overtly racial references that are to follow. 'This odious society' of the Jerusalem Jews, we are quickly informed, 'could not fail to weigh heavily on the tender and susceptible minds of the north'.[28] The Galileans, of whom Jesus was clearly one, differed from the Jews of Jerusalem in speech, temperament, and even race. 'It was believed (not without reason) that they were not of pure Jewish blood, and no one expected Galilee to produce a prophet.'[29] Having thus established that the nature-loving and mystical Galileans were not merely 'northerners', Protestant and anti-clerical in their leanings, but also of a distinctively different race from the legalistic fanatics of Jerusalem, the implicit conclusion should come as no surprise.

It was not Renan, however, but Émile Burnouf who spelled it out. His great work *La Science des Religions* had been published serially in Arnold's favourite magazine, the *Revue des deux mondes*, between 1864 and 1869, and appeared in book form in 1872.[30] Arnold makes use of it in *Literature and Dogma*. Burnouf has absorbed Gobineau and Renan, adding the information gleaned from Bunsen that there were among the Jews two distinct races – one black, the other merely dark. The Jews of Galilee were clearly Aryans. Jesus, the true Indo-European, had encountered his natural enemies in the Semites of Judea who had crucified him.

Lest we should feel any lingering doubts about the contrast between these New Testament Aryans and their Semitic Old Testament opponents, Burnouf explains that this is not just a matter of culture, but goes right to the very bone. For a late-twentieth-century audience, the details have a familiar and chilling ring:

> Those scholars who have studied anthropology almost all agree in placing the Semites between the Aryans and the yellow peoples: not that their distinctive traits betoken a medium condition between those of our race and those of eastern Asiatics; but notwithstanding their being far superior to the yellow race, they betray with regard to us such disparities as to prevent their being confounded with Indo-Europeans. A real Semite has smooth hair with curly ends, a strongly hooked nose, fleshy, projecting lips, massive extremities, thin calves, and flat feet. And what is more, he belongs to the occipital races; that is to say, those whose hinder part of the head is more developed than the front. His growth is very rapid, and at

fifteen or sixteen it is all over. At that age the divisions of his skull which contain the organs of intelligence are already joined, and in some cases even perfectly welded together. From that period the growth of the brain is arrested. In the Aryan races this phenomenon, or anything like it, never occurs, at any time of life, certainly not with the people of normal development. The internal organ is permitted to continue its evolution and transformations up till the very last day of life by means of the never-changing flexibility of the bones.[31]

To Arnold's credit he will have none of this. In *Literature and Dogma* he dismisses Burnouf with suitable ridicule.

Israel, therefore, instead of being a light to the Gentiles and a salvation to the ends of the earth, falls to a place in the world's religious history behind the Ayra. He is dismissed as ranking anthropologically between the Ayras and the yellow men; as having frizzled hair, thick lips, small calves, flat feet, and belonging, above all, to those 'occipital races' whose brain cannot grow above the age of sixteen; whereas the brain of a theological Ayra, such as one of our bishops, may go on growing all his life.[32]

But even if he was not prepared to go along with every line of Burnouf, he does not seem to have dissented much overall. Hellenism is of Indo-European origins, and is therefore native to the English tradition, whether it be seen as predominantly Teutonic, as Carlyle believed, or Celtic, as Renan fondly hoped; Hebraism is Semitic and therefore alien. *Culture and Anarchy*, for instance, is clear about rejecting the Mosaic law:

immense as is our debt to the Hebrew race and its genius, incomparable as is its authority on certain profoundly important sides of our human nature, worthy as it is to be described as having uttered, for those sides, the voice of the deepest necessities of our nature, the statutes of the divine and eternal order of things the law of God, – who, that is not manacled and hoodwinked by his Hebraism, can believe that, as to love and marriage, our reason and the necessities of our humanity have their true, sufficient, and divine law expressed for them by the voice of any Oriental and polygamous nation like the Hebrews? Who, I say, will believe when he really considers the matter, that where the feminine nature, the feminine ideal, and our relations to them, are brought into question, the delicate and apprehensive genius of the Indo European race, the race which invented the Muses, and chivalry, and the Madonna, is to find its last word on this question in the institutions of a Semitic people whose wisest king had seven hundred wives and three hundred concubines?[33]

Nevertheless, however adamant Arnold is that the book of Leviticus should not be allowed to legislate whether we be allowed to marry our deceased wife's sister, in what other respects the new scientific and Aryan creed might differ from its Semitic prototype is less than clear. The attempt of *Literature and Dogma* to substitute morality for religion, thus ensuring the superiority of dogma over literature, ends in self-contradiction.[34] Arnold, as always, is bedevilled by bursts of honesty that threaten to complicate or even undermine his whole carefully appropriated system. There are, he admits, many ways in which the English might even be considered Hebraic rather than Hellenistic. Renan himself had believed that there were parallels between the Protestant English and the ancient Jews. Arnold's prose is a masterpiece of balance – or indecision.

> Eminently Indo-European by its *humour*, by the power it shows, through this gift, of imaginatively acknowledging the multiform aspects of the problems of life, and of thus getting itself unfixed from its own over-certainty, of smiling at its own over-tenacity, our race has yet (and a great part of its strength lies here), in matters of practical life and moral conduct, a strong share of the assuredness, the tenacity, the intensity of the Hebrews.[35]

The real struggle is not between Hebrews and Hellenes, nor between Philistines and Barbarians – such polar opposites will always be present in any society – but whether the predominant culture will be Hebraic, narrow, rigid and moralizing, or Hellensitic, free, open and responsive. And in case we should be tempted to suppose that this is a conflict between religion and free-thought, his private notebooks illustrate how profoundly religious were his own concerns at this time. At the same time Arnold's correspondence reveals that his model Hellene, radiating 'sweetness and light', was none other than John Henry Newman.[36]

This astonishing admission illustrates perhaps better than anything else the real emptiness of Arnold's scientific cupboard. If ethnography seemed to offer a scientific basis for Schleiermacher's conviction that the Old Testament was no more than an appendix to the New, historically interesting but no practical guide to belief or conduct,[37] it was harder to imagine that Greek and Roman mores, Plato and Socrates included, would be any better than the Norse Valhalla or ancient Teutonic tree-worship in providing a serious replacement – and, to be fair, Arnold was well aware of the fact.

It is perhaps hardly surprising, given his cultural and religious concerns, that he should have been more interested in the biology of Gobineau than of Darwin. Yet Arnold's ethnography was more than simply a pseudo-science, a grand narrative that failed, a dead end in the history of ideas and a reminder of how alien the real substructure of even the most familiar of ideas can turn out to be. If we simply class it with the great failures in the history of science, with phlogiston, with Lamarckian evolution, or with the idea of the ether, we do it an injustice. The idea of a new humanistic Christianity, consonant with modern science and purged of its Semitic crudity and superstition, exerted a constant ideological pull not just on Carlyle, Kingsley and Arnold, but even on writers as different from them as George Eliot and Hardy. If with hindsight we can see in it less of 'sweetness and light' than the shadow of an ideology responsible for the death of six million Jews in the Nazi death-camps, and for more than forty-five million others, from Russia to North Africa, that is a mark at least of the importance for many of the quest. Whether Nazis or Marxists have been responsible for more untimely deaths this century is more a matter of definition than debate, but such are the Victorian origins of what was arguably, at least, the most powerful ideology of the twentieth century.

Notes

1. See David Newsome, *The Parting of Friends* (London: Murray, 1966), p. 289.
2. Geoffrey Faber, *Oxford Apostles* (London: Faber, 1933), p. 306.
3. Robert Lowth, *Isaiah: a New Translation* (1778), 5th edn, 2 vols (Edinburgh, 1807), I, lxx.
4. See Stephen Prickett, *Words and the Word: Language, Poetics and Biblical Interpretation* (Cambridge: Cambridge University Press, 1986) and François Deconinck-Brossard, 'England and France in the Eighteenth Century', in *Reading the Text: Biblical Criticism and Literary Theory*, ed. Stephen Prickett (Oxford: Blackwell, 1991).
5. Newsome, *The Parting of Friends*, p. 78.
6. Most notably Herbert Marsh, Lady Margaret Professor of Divinity 1807–16, who had translated Michaelis's *Introduction to the New Testament* into

78 *Rethinking Victorian Culture*

English with original notes of his own (1793–1801) and was responsible for the introduction of German scholarship in the University.
7. Adrian Desmond, *The Politics of Evolution* (Chicago: University of Chicago Press, 1989); James Moore, *The Post-Darwinian Controversies* (Cambridge: Cambridge University Press, 1979); Desmond and Moore, *Darwin* (London: Michael Joseph, 1991).
8. Edward Gibbon, *The Decline and Fall of the Roman Empire*, 6 vols (London: Everyman, 1910), V, 14–15.
9. Charles Kingsley, *Hypatia; or, New Foes with an Old Face* (London: Everyman, 1907), p. 11.
10. *Ibid.*, p. 226.
11. *Ibid.*, p. 9.
12. *Ibid.*, p. 7.
13. John Sutherland, ' "Two-timing novelists": W. M. Thackeray, *Pendennis*, Mrs Gaskell, *A Dark Night's Work*, and Antony Trollope, *Rachel Ray*', in *Is Heathcliff a Murderer? Great Puzzles in Nineteenth-Century Literature* (Oxford: Oxford University Press, 1996), pp. 123–38.
14. Mrs Humphry Ward, *Robert Elsmere* (1888), ed. Clyde de L. Ryals (Lincoln: University of Nebraska Press, 1967), pp. 276–7.
15. Gotthold Ephraim Lessing, *The Education of the Human Race* (1780); F. C. Schiller, *On the Aesthetic Education of Man* (1795), trans. R. Snell (London: Routledge, 1954).
16. *Complete Prose Works of Matthew Arnold*, ed. R. H. Super, 11 vols (Ann Arbor: University of Michigan Press, 1960–78), V, 173.
17. *Letters of Matthew Arnold 1848–88*, ed G. W. R. Russell, 2 vols (New York: Macmillan, 1900), I, 381.
18. *Prose Works*, p. 26.
19. F. E. Faverty, *Matthew Arnold the Ethnologist* (1951) (Evanston, IL: Northwestern University Press, 1968), p. 169.
20. Ernest Renan, *The Life of Jesus* (London: Watts, 1935), pp. 28–9.
21. *Ibid.*, p. 99.
22. *Ibid.*, p. 117
23. *Ibid.*
24. *Ibid.*, p. 121.
25. *Ibid.*, p. 125.
26. *Ibid.*, p. 64.
27. *Ibid.*, p. 124.
28. *Ibid.*, p. 118.
29. *Ibid.*
30. See Faverty, *Arnold the Ethnologist*, pp. 170–2.
31. Émile Burnouf, *The Science of Religions*, trans. Julie Leibe (London: Swann Sonnenschein, Lowrey, 1888), pp. 190–1.
32. Arnold, *Prose Works*, V, 161.
33. *Culture and Anarchy*, ed. Dover Wilson (Cambridge: Cambridge University Press, 1966), p. 184.
34. See Stephen Prickett, *Romanticism and Religion* (Cambridge: Cambridge University Press, 1976), Chapter 8, pp. 2–48.
35. *Ibid.*

36. David DeLaura, *Hebrew and Hellene in Victorian England* (Austin: University of Texas Press, 1969), pp. 61–81.
37. See for instance *The Christian Faith*, 2nd edn, trans. H. R. Mackintosh and J. S. Stewart (Edinburgh: T. & T. Clark, 1928), para. 12 (pp. 60–2).

6
Dickens and the Millennium
Andrew Sanders

> O ye hypocrites, ye can discern the face of the sky; but ye cannot discern the signs of the times.
>
> Matthew 16: 3

To most twentieth-century Victorianists the title 'Signs of the Times' has automatically suggested a reference to Carlyle's famous essay of that name, published in the *Edinburgh Review* in 1829. The same may not be true of scholarly Victorianists of the twenty-first century. Volume 52 of the *Nineteenth-Century Short-Title Catalogue* published in 1994, and covering the years 1816 to 1870, lists some eighty-one similar titles, most of them post-Carlylean and most of them warning of an imminent end of the world, or, at the very least, counselling an urgent need for repentance in the face of changing and challenging times.[1] They are all, in their often modest or clumsy way, indicators of an all too often overlooked aspect of nineteenth-century popular culture. These titles include the anonymous *The Signs and Duties of the Times: A Letter from a Country Clergyman to his Parishioners* (1821), *The Signs of the Times: or, A Glance at the present Social, Political and Religious Features of Society, with the Duty of the Citizen and Christian by Observator* (1848), *Signs of the Times: or, The Pope and Protestantism* (1850), *The Signs of the Times: Comprising a History of Spirit-Rapping in Cincinnati and Other Places* (1851), a verse tract of 1860 and a pamphlet ascribed to Carlyle's friend and fellow Scot, Edward Irving, and published in the same year as the now more celebrated essay. When we read in Damian Thompson's otherwise exemplary recent study of millenarianism, *The End of Time: Faith and Fear in the Shadow of the Millenium*, that 'to study the continuing influence of the Christian concept of End-time [since the eighteenth century], we must cross the Atlantic', we should in fact be wary of crossing the great divide too precipitously.[2] British millenarianism remained vigorously alive in the first half of the nineteenth century, and its proponents were as anxious as their brasher American counterparts to offer analyses of what Carlyle called the 'great outward changes' and the 'boundless grinding collision of the New with the Old'.[3]

There was nothing peculiarly 'Victorian' or even 'proto-Victorian' about the 'signs of the times' detected by the millenarianists of the 1830s and 1840s. Although reproductions of John Martin's lavish canvases of ancient cataclysms and the impending 'Great Day of His Wrath' gripped the imaginations of an Evangelically tutored generation (Martin mezzotints were, for example, hung conspicuously in Haworth parsonage), both the paintings and the interest they attracted reflected a long-standing and emphatically biblical fascination with divine wrath, human frailty and the end of time.[4] Nevertheless, as J. F. C. Harrison argues in his authoritative study, *The Second Coming: Popular Millenarianism, 1780–1850*, the political upheaval of the French Revolution had excited 'a spate of interpretations on both sides of the Atlantic to show that the world was entering upon the last days'.[5] For Harrison associative links can be forged between the private apocalyptic visions of William Blake and the phenomenal popular appeal of the prophesies of Joanna Southcott (who had attracted some 20,000 disciples by 1815).[6] As Charles Wesley's great advent hymn 'Lo He comes with clouds descending' also serves to suggest, a preoccupation with the Second Coming had been as much part of the Christian consciousness in the so-called 'age of enlightenment' as it was a central element in the supposedly more fevered religious imaginations of the post-Romantic age.

The opening decades of the nineteenth century had, however, witnessed an extraordinary burgeoning of millenarianism, both in terms of new sects and in more orthodox responses to traditional eschatological doctrine. Even such a High Church ritualist as the young Maria Rossetti could, thanks to the teaching of the Reverend William Dodsworth at Christ Church, Albany Street, in the early 1840s, express fear at looking at the Egyptian mummies in the British Museum lest the General Resurrection should suddenly reanimate the occupants of the glass cases.[7] Dodsworth (1798–1861) was to become a prominent convert to Roman Catholicism in 1851. The popular appeal of early-nineteenth-century millenarianism was evidently classless and genderless and it crossed boundaries of education and Christian affiliation. A group of predominantly upper-middle-class Anglicans (two of them priests), encouraged by the organizational talents of the banker Henry Drummond, formed the apostolic nucleus of the Catholic Apostolic Church in the early 1830s. These disciples of the magnetic Scots preacher Edward Irving (1792–1834), followed their spiritual mentor in insisting on the imminence of the Second Coming and in affirming that:

> the present Christian Dispensation would not pass insensibly into the Millenial state by gradual increase of the preaching of the Gospel. It would be terminated by judgements ending in the destruction of the visible Church in the same manner as that in which the Jewish Dispensation had been terminated.

The Irvingites argued that a period of 1260 years which had commenced with the reign of Justinian had come to an end with the extinction of the French monarchy in 1792 and 'that the vials of the Apocalypse began then to be poured out'.[8] Their doctrine, and their urgent plea for rigorous repentance and religious renewal in preparation for Christ's Advent, was echoed in Scotland by the Reverend Alexander Keith, the Minister of St Cyrus, Kincardineshire. Keith, a Doctor of Divinity of Aberdeen University, first published his enduringly influential *The Signs of the Times, as denoted by the Fulfilment of Historical Predictions, traced down from the Babylonish Captivity to the Present Time* in Edinburgh in two volumes in 1832. It is a scrupulously laboured, but nonetheless urgent, attempt to locate the nature of the coming Armageddon. Keith opened his study with the bold prophetic warning:

> Never, perhaps, in the history of man, were the times more ominous, or pregnant with greater events than the present. The signs of them are in many respects set before the eyes of men, and need not to be told; and they strike the senses so forcibly, and come so closely to the apprehension of all, that they may be said to be felt, as well as to be seen. The face of the sky never indicated more clearly an approaching tempest, than the signs of the times betoken an approaching convulsion, – not partial, but universal.[9]

He accompanied his survey of the manifestations of the approaching cataclysm with maps and diagrams, spurred into intellectual action not simply by the fall of the French monarchy and the current unsteadiness of the Pope's throne (Rome is self-evidently the 'seat of the Beast') but also by the decaying and turbulent state of the Ottoman Empire (it would naturally be in Palestine that Christ would first remanifest himself). Keith's compelling study appears to have found a receptive audience. It had reached its eighth edition by 1847, by which time the author had joined the new Free Church of Scotland and had turned himself into a pioneer exploiter of the daguerreotype in order to further explain the landscapes of the Holy Land in which the divine future would be realized.[10]

Most early-nineteenth-century millenarianists and their followers were far less sophisticated than Alexander Keith. These schismatics were often ill-educated and of lower-middle- or working-class stock. It should also be stressed that their message had quite as forceful an appeal to the rapidly expanding, disorientated urban proletariat and to the often distressed agrarian population as did the political (and notably Chartist) ideas which have since almost exclusively engaged the interest of historians. The sometime farmer and collier John Wroe (1782–1863), who attracted a substantial following in Lancashire in the 1820s, had convinced himself of the truth of Christian Israelitism and had insisted that his male disciples adopt beards, circumcision and strict dietary laws. In order to prepare for the Second Coming of Christ, Wroe had attempted to transform Ashton-under-Lyne into the New Jerusalem, with a sanctuary in Church Street and with four 'Gates of the Temple' situated at cardinal points on the outskirts of the town. When the Second Coming failed to materialize at the predicted time, many of Wroe's Israelites were absorbed into the equally urgent millenarianist new creed of Mormonism.[11] The largely uneducated Irishman John Ward (1781–1837), who had trained as a shipwright in Bristol and as a shoemaker in London, shifted from Calvinism to Methodism, and from the Baptists to the Sandemanians, before discovering his neo-Southcottian vocation in Walworth in 1827 as 'Zion' or 'Shiloh', the Prince of Peace, and announcing his apocalyptic destiny to large crowds in the north of England.[12]

It was, nevertheless, in early-nineteenth-century America that millenarianist doctrines took deepest root. Most sensational and successful, in the short term at least, was the New York farmer William Miller (1782–1849), who confidently predicted that the end of the world would fall some time between 21 March 1843 and 21 March 1844 and who propagated his ideas through a newspaper inevitably entitled *Signs of the Times*. If many of the disappointed thousands of Millerites abandoned their new-found faith when March 1844 passed without incident, others desperately sought answers in a revised chronology, earnestly waiting on hillsides for the Lord's triumphant descent in dreadful majesty.[13] Far more enduring, if equally inclined to resort to revised chronologies, have been those still-burgeoning churches of the Seventh-Day Adventists and the Latter-Day Saints (the Mormons). The former may well claim Millerite roots in the 1830s, although the Adventist faith tended to define itself clearly only in the 1860s and particularly (in British terms) after a successful mission to Southampton in 1878.[14] The impact of the Mormons on Britain was far more immediate and fruitful. Joseph Smith received his first vision of an angel in 1823, an angel who proclaimed 'that the preparatory work for the second coming of the Messiah was speedily to commence: that the time was at hand, for the Gospel in all its fullness to be preached in power, into all nations, that a people might be prepared for the millennial reign'. The 'Book of Mormon' was published in 1831, the year after the Church of the Latter-Day Saints had been founded in Manchester, New York. The details of Smith's often offensively novel doctrines, of his murder in 1844 and of the great Mormon trek to Salt Lake City, Utah, under the leadership of Brigham Young in 1846–7, need not be rehearsed here. What was significant for the purposes of this essay were the Mormon missions to England (commencing in June 1837) by means of which British working-class millenarianism became integral to an American mainstream. By 1840 there were some 4000 English Saints; by 1843 7500; by 1851 33,000. Between 1837 and 1851 some 17,000 had emigrated to the United States en route for the refuge on the Great Salt Lake where the elect were to await the Second Coming of the Lord. These emigrants were, according to a Liverpool shipping manager, 'principally farmers and mechanics, some few clerks [...] many highly respectable'.[15]

The subject of Mormon mission and Mormon emigration leads us at last to Dickens, for his fascination with the Latter-Day Saints gave rise to one of the most remarkable of the 'Uncommercial Traveller' essays ('Bound for the Great Salt Lake', in *All the Year Round,* 4 (July 1863)). From the evidence of his fiction, Dickens would appear to have been singularly unresponsive to millenarianist doctrine and not in the least persuaded by its urgency in the 1830s and 1840s. Perhaps the most notable millenarianist in his fiction is the Reverend Melchisedek Howler in *Dombey and Son* (1846–8). Howler is introduced in Chapter 15, without much evident respect, as Mrs MacStinger's spiritual mentor. Howler's religious history is brusquely and satirically delineated. He was, we are told, one day discharged from the West India Docks on a false suspicion (got up expressly against him by the General Enemy) of screwing gimlets into puncheons, and applying his lips to the orifice, had announced the destruction of the world for that day two years, at ten in the morning, and opened a front parlour for the reception of ladies and gentlemen of the Ranting persuasion, upon whom, on the first occasion of their assemblage, the admonitions of the Reverend Melchisedech had produced so powerful an effect, that, in their rapturous performance of a sacred jig, which closed the service, the whole flock broke through into a kitchen below, and disabled a mangle belonging to one of the fold.[16]

Howler is therefore something of a John Ward, shot through with much of the Dickensian comic distaste for Dissenting ministers which had earlier shaped the characterization of Stiggins in *Pickwick Papers* (1836–7). Howler has only a marginal role in the social and fictional fabric of *Dombey and Son*, but he has one key function as the novel draws to its benign conclusion: he consents 'on very urgent solicitation, to give the world another two years of existence, but had informed his followers that, then, it must positively go' (Chapter 60, p. 814). He is therefore joined, as grudgingly and half-heartedly as a Shakespearian melancholic, to the spirit of general rejoicing which marks the tone of the novel's last chapters. Despite his revised chronology, Howler's preaching of an imminent end is, in fictional terms, superseded by human events.

Melchisedech Howler is a comic device, almost a throwaway one. More serious, and far more dangerous in Dickensian terms, are other false prophets, notably those who haunt the opening pages of that self-consciously 'prophetic' novel, *A Tale of Two Cities* (1859). The 'noisiest authorities' who, at the end of the first paragraph insist that the period be 'received, for good or for evil, in the superlative degree of comparison only' are as likely to be eighteenth-century millenarianists as they are Enlightenment *philosophes* and theoretical apologists for the *status quo*. If 1775 is 'so far like the present period' that original readers of the novel might have felt obliged to reach out for modern parallels, Dickens makes plain in his third paragraph that the England of the late eighteenth century was peculiarly susceptible to the apocalyptic warnings of false prophets. Joanna Southcott is mentioned by name (though we are reminded that her first prophecies date only from 1792). Dickens then proceeds to conflate two references to the mass hysteria attendant on millenarian warnings, both of them damp squibs. His 'prophetic private in the Life Guards' who has 'heralded the sublime appearance by announcing that arrangements were made for the swallowing up of London and Westminster', refers back to an incident of 1750. This incident is abruptly linked to 'the reveries of a crazy prophet', mentioned in the *Annual Register* for 1775, who had announced that Deptford and Greenwich 'were to be swallowed up by an earthquake'.[17] What Dickens's narrator is doing in the early pages of *A Tale of Two Cities* is attempting to distinguish between false prophets of doom, preoccupied with the unknown and unknowable, and the real 'signs of the times', the warnings against corruption, privilege and misgovernment which appear to be writ large in contemporary society. The cataclysm of the French Revolution, he implies, is already implicit in the nature of the society of the *ancien régime*, while warnings of an imminent Second Coming are a distraction from more pressing human and divine concerns.[18] This threat of a decidedly secular apocalypse parallels that which runs through *Bleak House* (1852–3), a novel much concerned with last things, with ends of term, judgement days, signs and tokens and with variously soggy, muddy and combustive ends. The darkness which pervades Dickens's novels of the late 1850s is that of a polluted and polluting society, not of a storm cloud foreshadowing the millennium. In terms of his fiction, the apocalypse is an imminent and human event, rather than a far-off and divine one.

When, however, in 1863 Dickens, as the 'Uncommercial Traveller for the firm of Human Interest Brothers', encounters a substantial group of Mormon millenarianists on the *Amazon,* an emigrant ship moored in the Thames, his tone is quizzical rather than ironic. The Uncommercial is evidently anxious to score intellectual points off the men he questions, but he systematically fails to hit home. His pose of a sceptical unbeliever is, however, carefully maintained. When he debates with the Mormon Agent on the *Amazon* he admits to be 'burning to get at the prophet Joe Smith' but his ardour is deflected by an ambiguous answer to a provocative question. A Wiltshire labourer, recently converted to Mormonism, creates a similar discomfiture in his interlocutor. When the Uncommercial opens a Mormon hymn-book he insists that it proved to be 'by no means explanatory to myself of the new and Everlasting Covenant, and not at all making my heart an understanding one on the subject of that mystery'. At the conclusion of the essay he still feels obliged to press his scepticism on his reader. The Mormon emigrants may be singularly disciplined and well-behaved but, he insists: 'What is in store for the poor people on the shores of the Great Salt Lake, what happy delusions they are labouring under now, on what miserable blindness their eyes may be opened then, I do not pretend to say.'[19]

This is the kind of provocative prophetic utterance which abstains from effective prophesy. The Mormons are deluded, it seems to say, but who am I to attempt to disabuse them of their folly? This group of 800 tidy, ordered, hymn-singing, tea-drinking pilgrims is in earnest, but their earnestness is not that of the Uncommercial or of Dickens. The 'great house of Human Interest Brothers' is self-evidently not much interested in dealing in popular Millenarianism. What remains remarkable about 'Bound for the Great Salt Lake' is the pervasive respect for respectability that it conveys. Because the likes of Melchisedech Howler are not present among the temperate, self-disciplined emigrants, there is quite simply no room for satire, let alone comedy.[20]

Dickens's response to the millenarianism which so marked the England of his boyhood and manhood would seem to have resembled that of Carlyle. When Carlyle selected the eminently potent title 'Signs of the Times' for his essay of 1829, he was substantially eschewing the current apocalyptic ramifications of the phrase. His essay is addressed to the secular and not to the divine, to the present state of society and not to an imminent spiritual crisis, to a new definition of culture and not to an explication of the Last Things. If Dickens was more inclined to jest at the millenarian insistence that time was coming to an end than he was to look up expectantly at the lowering doom clouds, it was because he was inclined to share much of Carlyle's secularism. To both men the crisis of the nineteenth-century 'Condition of England' seemed to have been humanly engineered and not divinely ordained. Nevertheless, to Carlyle at his most rhetorical and to Dickens at his most moral, the gradual redemption of modern society was precisely the task that God had entrusted to each single element of his human creation as long as God's gift of measured time endured. As Carlyle insisted at the conclusion to 'Signs of the Times',

> On the whole, as this wondrous planet, Earth, is journeying with its fellows through infinite Space, so are the wondrous destinies embarked on it journeying through infinite Time, under a higher guidance than ours [...]. Go where it will, the deep HEAVEN will be around it. Therein let us have hope and sure faith. To reform a world, to reform a nation, no wise man will undertake; and all but foolish men know, that only the solid, though a far slower reformation, is what each begins and perfects on *himself*.[21]

As the twentieth century draws to its close and a second millennium looms with all its renewed potential for the fulfilment of ancient apocalyptic prophecy, millenarianists from Texas to Japan have discovered a newly receptive pool of willing and sometimes suicidal disciples.[22] If press reports are to be believed, the Mormon church in Britain has grown thirty-fold in the past thirty years and will soon have as many members as the mainstream churches.[23] Dickens's fiction might ostensibly seem to be irrelevant to an exploration of the impending end of time. As an attempt to interpret the 'great outward changes', the secular 'signs of the times', many of them changes and signs closely related to those that so bemused and befuddled nineteenth-century millenarianists, his fiction still possesses a particular immediacy to those of us attempting to disentangle the social, moral and intellectual milieu of the Victorians.

Dickens seems to me the first truly 'modern' writer, in the sense that he was the first major writer to address the modern, industrialized, disconcerted urban condition, what the prophetic Carlyle had called 'the boundless grinding collision of the New with the Old'.[24] Now that 'modernity' itself has been (we are told) refigured as 'post-modernity', and as both modernities stumble blindly into new collisions and into twenty-first century marginality (or perhaps even obsolescence), it is possible that Dickens's relevance to the future will seem purely historical. But then again, if our world society is spared the threatened apocalypse, his work will perhaps continue to address the central problem that seems set to beset the twenty-first century: that of the innate untidiness and confusion of urban existence.

Notes

1. *Nineteenth-Century Short-Title Catalogue*, Series II Phase I, 1816–1870 (1994), p. 52.
2. Damian Thompson, *The End of Time: Faith and Fear in the Shadow of the Millenium* (London: Sinclair-Stevenson, 1996), p. 95.
3. 'Signs of the Times' (*Edinburgh Review*, 98 (1829)), repr. in *Scottish and Other Miscellanies*, intro. by James Russell Lowell (London: Everyman, 1967), p. 245.
4. For Martin and his impact on nineteenth-century England see William Feaver, *The Art of John Martin* (Oxford: Clarendon Press, 1975). Feaver quotes Bulwer-Lytton's declaration in *England and the English* (1833) that Martin was 'the greatest, most lofty, the most permanent, the most original genius of his age' (p. 141).
5. (London: Routledge & Kegan Paul, 1979), pp. 5, 57.

6. Harrison, *The Second Coming*. pp. 81, 86–134.
7. Cited by Jan Marsh in her *Christina Rossetti: a Literary Biography* (London: Cape, 1994), p. 57.
8. For Drummond and the Catholic Apostolic Church see P. E. Shaw, *The Catholic Apostolic Church* (New York: King's Crown Press, 1946) and R. E. Davenport, *Albury Apostles: the Story of the Body known as the Catholic Apostolic Church*, 2nd edn (London: Catholic Apostolic Church Library, 1970; 1975). See also [anon.], *'The Years of Ferment': Being the Story behind the Building of the Catholic Apostolic Church in Albury, Surrey in 1840* (Albury Park: [privately printed], 1980). For Irving see Margaret Oliphant, *The Life of Edward Irving*, 2 vols (London: Hurst and Blackett, 1862).
9. Alexander Keith, S*igns of the Times*, 3rd edn, 2 vols (Edinburgh: William Whyte, 1833), I, 1.
10. Keith travelled to Palestine in 1844 (information from the *Dictionary of National Biography*). Keith's interest in photography was probably fostered by his acquaintance with the pioneer Scottish photographers David Octavius Hill (1802–70) and Robert Adamson (1821–48). Hill and Adamson took a calotype of Keith in the 1840s which was included in the centenary exhibition of their work held by the Scottish Arts Council in Edinburgh in 1970.
11. For Wroe see Harrison, *The Second Coming*, pp. 138–48.
12. For Ward see Harrison, *The Second Coming*, p. 230. See also *Dictionary of National Biography*.
13. For Miller and the Millerites see Harrison, *The Second Coming*, p. 192. See also Thompson, *The End of Time*, pp. 98–101.
14. For the Adventists, see the entry and bibliography in F. L. Cross and E. A. Livingstone (eds), *The Oxford Dictionary of the Christian Church*, 2nd edn (Oxford: Oxford University Press, 1974). See also Thompson, *The End of Time*, pp. 282–4.
15. See Harrison, *The Second Coming*, pp. 176–190.
16. *Dombey and Son*, World's Classics (Oxford: Oxford University Press, 1982), Chapter 15, p. 206.
17. For these references in full see Andrew Sanders, *The Companion to 'A Tale of Two Cities'* (London: Unwin Hyman, 1988), p. 29.
18. The 'millenarianist' mood of the 1790s has recently been explored by Roy Porter in his 'Visions of Unsullied Bliss' in Asa Briggs and Daniel Snowman (eds), *Fins de Siècle: How Centuries End, 1400–2000* (New Haven: Yale University Press, 1996), pp. 125–55.
19. 'Bound for the Great Salt Lake', repr. in *The Uncommercial Traveller and Reprinted Pieces*, The Oxford Illustrated Dickens (Oxford: Oxford University Press, 1968), p. 232.
20. For the factual basis to the essay see Richard J. Dunn, 'The Unnoticed Uncommercial Traveller', *Dickensian*, 64 (1968), 103–4.
21. 'Signs of the Times', *Edinburgh Review*, 98 (1829); repr. in *Scottish and other Miscellanies*, p. 245.
22. For the new Millenarianists see Damian Thompson *The End of Time*. For the impact of millenarian thought on twentieth-century totalitarianism, see Norman Cohn, *The Pursuit of the Millenium: Revolutionary Millenarians and Mystical Anarchists of the Middle Ages*, new edn (London: Pimlico, 1993). See

90 *Rethinking Victorian Culture*

also Asa Briggs, 'The 1990s, The Final Chapter' in Briggs and Snowman (eds), *Fins de Siècle*, pp. 197–233.
23. 'Mormons catching up with mainstream churches', *The Times*, 26 February 1996, p. 8.
24. 'Signs of the Times', *Scottish and other Miscellanies*, p. 245.

7
Re-Arranging the Year: the Almanac, the Day Book and the Year Book as Popular Literary Forms, 1789–1860

Brian Maidment

1.

In 1789 William Godwin wrote in *The Annual Register* (itself an important example of the year-book genre): 'From here we are to date a long series of years, in which France and the whole human race are to enter into possession of their liberties'.[1] This essay concerns some of the literary and figurative ways in which succeeding generations sought to 'enter into possession' of 'a long series of years' by reformulating and reconfiguring those years as an aspect of their 'liberties' – that is, by literally and figuratively taking liberties with the calendar. These liberties included the need and opportunity (perhaps even the assertion of a right) to re-make traditional calendars – be they ecclesiastical, agricultural or social – in ways which reflected, subverted or even tried to reconstruct the experience of cyclical time as a form of social practice. During the years when, under the regime of early industrialism, time was being formulated as clock time or shift time, and when the legacy of the revolutionary making and remaking of time schemes as expressions of ideology and political purpose remained fresh in people's minds, it was almost inevitable that definitions of the calendar should have been competed for across the range of political and social opinion. It was in the genres and forms of popular culture that these contests were most likely to manifest themselves, especially in those genres (the almanac, the day book, and the year book) which traditionally had formulated calendar time as a structure (itself both cycle and sequence) of social and intellectual experiences.

This essay seeks to be exploratory rather than definitive in its necessarily simple narrative of the calendar as a literary and ideological form. Few of the literary forms of the calendar show the spectacular political purposefulness or linguistic and emblematic drive of the revolutionary calendar in France, nonetheless they do suggest how important and sustained a conflict over the significance of dates existed within an increasingly industrial Britain between the French Revolution and the establishment of a Victorian equipoise – the term 'dates', of course, implies a combination of remembered significance and immediate social activity or ceremonial. Thus one of the arguments of this essay is that there was a radical almanac tradition in Britain which manifested itself crucially in the 1830s and 1840s. However, the fervour and energy figured in the revolutionary calendar began to fade in the early Victorian period, perhaps another symbolic casualty of the collective failures of revolutionary movements across Europe in 1848, and by mid-Victorian times, the almanac and year book had, in popular culture at least, lost all their significant histories – their Georgic connection with the agricultural year was almost as diluted as their distant recall of the demands of the liturgical year. Indeed, it was the comic almanac, which in the hands of George Cruikshank, *Punch*, and their successors, became an almost parodic reminder of lost potentialities and histories, that dominated the early Victorian period and survived, in various popular forms, into the twentieth century. However acute early Victorian comic draughtsmanship may be as a form of social commentary, these almanacs essentially celebrated social pleasure as both street theatre and domestic experience through structures which derived from, but did not substantially invoke, the devotional cycle of the ecclesiastical and liturgical year. But before reaching the untroubled celebratory pleasures of the mid-century, it is worth a glance at the more contested and controversial conceptions of time, dates, remembrance and social ceremony which are implicit in the various genres of annotated calendars which I have chosen to discuss here.

2.

Most histories of popular literature have little if anything to say about almanacs as a distinct literary genre – Victor Neuburg's *Popular Literature*, for example, has a single illustration and passing reference to sixteenth- and seventeenth-century almanacs[2] – although most acknowledge their widespread existence and allude to their importance to the popular imagination.³ But three books which discuss wider themes in the history of popular consciousness offer brief but hugely suggestive comments. Louis James's *Print and the People, 1819–1851* gives considerable space to illustrations of early-nineteenth-century almanacs, and suggests their importance in maintaining a predictive tradition which underpins the survival of popular superstition.⁴ The attack on precisely those elements of popular superstition, of course, forms part of the grand narrative of Keith Thomas's *Religion and the Decline of Magic* which describes the social contest between organized propagandist religion and the popular will to superstition, credulousness and wonder in the early modern period and beyond.⁵ In a different context, David Vincent's work on artisan culture links almanacs in to the rise of advertising and the penny post in the early Victorian period.⁶

But if James, Thomas and Vincent only provide important insights, almanacs do now have a proper continuous scholarly commentary which describes their history from the Renaissance through into the Victorian period. Bernard Capp's *Astrology and the Popular Press: English Almanacs, 1500–1800* defines the central relationship between almanacs and popular belief in the early modern period.⁷ While there is an obvious slackening of this connection between superstition, prediction and the almanac form in the Victorian period, it would nonetheless be possible to trace the survival of the kinds of literature described by Capp through its later history using examples of the kind printed by Louis James. More useful still might be a description of the literature developed to expound the popular devotional year, a literature most famously represented by Keble's *Christian Year* which was first published in 1827. But the aim of this essay is to trace a trajectory for the almanac which is essentially secular, politically aware, and aimed, up to the 1840s at least, at that fiercely contested section of the reading public, the emergent, self-educated, culturally ambitious artisan classes. And in discussing these kinds of readers, it is essential also to bear in mind the revolutionary tradition of the calendar, and the way in which almanacs were used to represent the political and ideological struggles in France. Lise Andries's essay, 'Almanacs: Revolutionizing a Traditional Genre' (1989), describes the political recognition in revolutionary France that the organization of the year into months, days and hours is not just a social convenience but also the manifestation of power structures within the culture: the 'ecclesiastical' year with all its hierarchies, obligations and anniversaries; the 'feudal' year which built a calendar out of the demands of agricultural production, and so on.[8] The attempts to re-formulate the year using new terms based on 'seasons' is one obvious way in which revolutionary politicians used language and emblems as a political discourse. Andries describes the ways in which almanacs were used as part of this ideological struggle, and if nothing so dramatic can be found in Britain, nonetheless the emblematic reformism of, say, the Owenite almanacs (see Figure 2) or the revolutionary rhetoric of *The Reformers' Almanac* show that the French lesson had been learnt in early industrial Britain.[9] Capp's and Andries's work, coupled with considerable academic interest in calendars, most substantially represented by Ronald Hutton's recent *The Stations of the Sun: A History of the Ritual Year in Britain*, gives an early history and context for a long tradition of almanac production which needs to be brought to bear on nineteenth-century developments in form, readership and ideology.[10]

Maureen Perkins's book *Visions of the Future*, as well as providing a scholarly account of the almanac genre in the late-eighteenth and early-nineteenth centuries, also offers scholars an informed place for disagreement.[11] Perkins provides a number of key contexts: she shows the way in which the Stationers' Company monopoly was exploited and challenged; she links the rationalist attacks on traditional predictive almanacs to Charles Knight and the Society for the Diffusion of Useful Knowledge; and she gives a fascinating account of the importance of almanacs abroad, especially in her case study of Australian almanacs. Yet there are still alternative narratives to be constructed. To my mind, three areas of Perkins's argument need reassessment. First, she links the 'rational' almanac tradition too closely to the SDUK without acknowledging the centrality of the almanac to other less institutional versions of self-education, like those represented by Dionysius Lardner's books, or even the five-volume *Instructor* published by John Parker for the SPCK (see Figure 4), both of which stressed the centrality of the almanac and of the calendar to artisan identity.[12] Second, she also under-represents the truly radical tradition of almanacs which was sustained and developed in Britain by such figures as William Hone and Henry Hetherington.[13] Third, the crucial role of illustration to the development of the almanac form is not given adequate attention.

The following essay argues that 'predictiveness' was not necessarily the only – or even the main – function of popular almanacs, and seeks to show a wide variety of ways in which they may be read as formative of popular attitudes at a moment in the 1830s and 1840s when the struggle for artisan cultural identity and tradition was at its most intense.

3.

For the purposes of this essay, I have defined almanacs, year books and day books as those literary genres predominantly structured by a wish to offer information, advice and commentary which is ordered by its day-by-day or seasonal relevance. This definition is contentious in so far as it confronts 'predictiveness' as the defining characteristic of the almanac genre. Using this definition, there are within these kinds of literature deep-seated, but ambiguous, assumptions about how these texts were intended to be read and used. While the reader is not prohibited from browsing in a disorderly way (and some day books, especially those published by Hone and Chambers, actively encourage this kind of miscellaneous reading), the implicit assumption is that a reader will read an almanac sequentially, and that the act of reading will be rendered appropriate and relevant in some manner to daily conduct through the coincidence of dates. This idea of the coincidence of dates, whatever superstitious or psycho-pathological depths it may touch, combines memorializing with acute consciousness of season, occasion and present activity. Anniversaries act both in the present and the past, and almanacs see both remembrance and the urgencies of day-to-day current activities as necessary to each other. Thus, understanding quite how almanacs informed the present conduct of their readers is a subject for considerable speculation: was their primary function *predictive*, like some aged predecessor of fashionable contemporary interest in daily prediction expressed by use of the *I Ching*, or had the old pre-literate dependence on superstition given way by the 1790s to more rational reminders of the seasonal obligations of the agricultural year? Were almanacs by this time essentially ideological in trying to formulate particular traditions of remembrance and daily activity in order deliberately to try to banish the predictive superstition of an earlier popular culture? There is no doubt that almanacs became institutionalized early in the nineteenth century just as legislative reliance on the Stationers' Company monopoly was breaking down. The Society for the Propagation of Christian Knowledge produced its own influential (and highly theorized) version as part of its educational programme, primarily defined by the five-volume and much reprinted *Instructor*. The Society for the Diffusion of Useful Knowledge followed suit (see Figure 3), with implications studied in detail by Perkins.[14]

But whatever the intentions of the organizations, entrepreneurs and ideologues who sponsored their production, it is possible that almanacs remained, to their readers at least, capable of a variety of interpretations which combined at will, or at need, utility, remembrance, entertainment and belief. It may be precisely this mixture of 'instruction and delight', drawing on irrational fears and beliefs as well as social utility, which allowed the almanac to survive as a socially contested but widely significant form, until its appropriation as a diversionary medium by some of the great entrepreneurs of Victorian bourgeois entertainment in the 1840s. For the Victorians, the almanac needed to become either harmlessly entertaining or else socially purposive in ways important for temperance campaigners, educators and advertisers alike.

Some of these diversities and contradictions show formally within the almanacs themselves. It is much easier to find pamphlet- or tract-type almanacs than broadside ones intended for public display. But my instinct, together with secondary information, suggests that the majority of almanacs were in fact in broadsheet format and intended for public display. The 'typicality' and 'representativeness' of surviving literature is an obvious issue for historians of popular culture. Given that broadsheet almanacs were intended to survive for a year only in the competitive, rough and tumble places of public display, it is hardly surprising that the more robust pamphlet- and chapbook-format almanacs have survived better. While major collections of ephemera like the John Johnson collection at Oxford, which formed the source for much of Louis James's *Print and the People*, contain examples of single-sheet almanacs, I have certainly found it hard to identify enough survivors to suggest the extent, variety and contexts of the broadside format.

Nonetheless, some perhaps typical examples of almanacs intended for public display can be found. One anonymous extra illustration to an early edition of *Pickwick Papers* shows, entirely without self-consciousness, an almanac hung prominently on an office wall. (This is an image which makes an obvious contrast with Hogarth's fourth plate to his sequence *Industry and Idleness* (1747), which displays an orderly loom shop with an emblematic and 'improving' almanac ostentatiously pinned to the master's desk in the foreground of the image.)[15] While almanacs must have been designed primarily to be functional in these kinds of setting, the display almanacs I have seen are all self-consciously decorative and suggest that careful consideration of their visual organization has taken place. One interesting and long-standing almanac comes to mind here – *The Oxford Almanac*, which was

Figure 1: Extra illustration to *Pickwick Papers* (1837)

produced annually with the same ostensible purpose as the dreary annual staff handbooks and calendars which I suspect teachers all use in their day-to-day lives.[16] *The Oxford Almanac* does indeed include term dates and examination periods, but the information is brilliantly integrated into a single, large, copper-engraved sheet which has as its central focus one or more superb views of the University buildings, views which were commissioned from the outstanding engravers of the day. The result is a sequence of some of the most accomplished and spectacular engravings of the eighteenth and early nineteenth centuries, which extend the limits of copper engraving considerably because of the amount of 'useful' text which the engravers were asked to incorporate into their designs. It would hardly be fair to call these almanacs simply 'functional'. Similarly, a rare survivor from the many Owenite almanacs of the 1830s, now in the Museum of Labour History in Manchester, which was intended to be hung in factory or workshop, combines wood engraving with commemorative dates in labour history and exhortatory prose and sayings, to make up a complex single display image.[17]

This carefully composed sheet is something both more attractive and more useful than the usual 'Thou shalt not' notices or lists of fines for misconduct. Furthermore, this Owenite almanac invokes history to define present purpose, and declares a tradition of honourable labour for public consideration and, no doubt, for the edification of the workforce. C. J. Grant's *Comic Almanac for 1836* offers a crude but effective way of maintaining an outspoken caricature tradition of satiric dissent through building the sheet out of a multiplicity of small lithographed images.[18] The resulting form is experimental, not just in terms of the almanac, but also as a way of organizing radical satirical images into a visually related single image, irrespective of their diverse subject matter.

Most of the surviving almanacs, however, were published as pamphlets or chap-books, with calendar information laid out in a series of grids which had become established by precedent as standard forms. The organization of complex information into tabular graphic form was obviously a central function of all almanacs from the predictive to the comic. But beyond this central function of providing data, early Victorian almanacs offered a huge variety of circumambient text which might include maxims and sayings, antiquarian accounts of customs and events, poems, scientific or pseudo-scientific information (especially concerning natural history), and even political polemic. In this period, many almanacs, even those produced commercially rather than institutionally, sought to deny the traditional association between superstition and the almanac – what one might call 'The Old Moore Connection' – in order to create a genre which was deeply rationalist, and which sought to construct a world out of fact, tradition and utility rather than out of superstition and prediction. Such a secular, rationalist and apparently progressive agenda was inevitably an ideological one, and it is no coincidence that the most celebrated proponents of the almanac and the year book were precisely those cultural entrepreneurs and organizations which had acknowledged – and exploited – the new discourses of popular progress and artisan self-cultivation

Figure 2: Social Reformers' Almanac for 1841

opened up by *The Rights of Man* (1791-2): William Hone, Mary and William Howitt, the SPCK, the SDUK – with the constant presence of its great innovating publisher Charles Knight – and, later, the Chambers Brothers.[19] Until the brilliant intervention of George

Cruikshank in the 1830s, with his continuing and astonishing power to reinscribe the anxiously demotic as the uncomfortably bourgeois, innovations and re-inventions within the almanac format were led by those cultural entrepreneurs who had begun to resolve through their publications the pressing issue of how literary and literate working men and women should become. The almanac was clearly an important piece of military equipment on 'the march of intellect'.

4.

How, then, given this complex and overlapping set of possibilities, is it possible to characterize the early Victorian almanac? While acknowledging the survival of predictive almanacs, which met a continuing demand for popular prophecy, the most obvious feature is the increasingly widespread production of 'useful' almanacs. These kinds of almanacs, usually in pamphlet form, listed hours of daylight, tides, phases of the moon and temperature ranges alongside information about major livestock markets and the liturgical calendar. Following these elements, which were presented in a tabular form, were less 'predictive' kinds of information: lists of Members of Parliament, details of the composition of public bodies, and an account of voting rights and the tax system. It is perhaps hard for us to recapture much sense of how this information might have been used on an everyday basis, let alone why the almanac apologists offered such impassioned accounts of its importance. Here is Dionysius Lardner writing in 1855: 'Of all books the Almanack is the most indispensable. So constant is the need for it, that, unlike other books, it is not deposited on the shelf, but lies ready at hand on the table'.[20] Lardner, in insisting on the 'constant and general utility' of the almanac, goes on to make the interesting point that almanacs had been subjected to Stamp Duty because of their presumed 'topical' content.

Lardner's rage at the 'fatal visitation of the Stamp Office' here is instructive because it places the almanac firmly at the centre of early Victorian debates about useful knowledge, and shows that government agencies at least regarded the almanac as something which required legislation. Given the contents of a 'useful' almanac sketched above, this may seem a curious response, as information about tides and saints' days may seem entirely value-free. The theorists and propagandists of 'useful knowledge', however, were well aware of how contentious and significant a form the almanac was, and of how the description of time and season was as much an ideological as a pragmatic act. For all their apparent neutrality, the 'useful knowledge' almanacs of the early Victorian period repressed some potential versions of the calendar and constructed new meanings of their own. The major act of repression has already been suggested as the implicit denial of any possible connection between the almanac year and all forms of popular superstition and prophecy. The 'useful' calendar had to be rational, utilitarian and aimed at supporting productive activity. As well as denying irrational prediction, the 'useful' almanac had also to offer a version of history (those things worth memorializing and marking through daily ritual) and a functionality which acknowledged Christian worship even as it supported secular commercial effectiveness. Given such important ideological purposes, it is hardly surprising to find extensive theorizing of the history and function of both calendar and almanac within many of the key self-education text-books of the early Victorian period: the SPCK's *Instructor* (which was extensively reprinted throughout the 1830s), Lardner's *Museum* (1855) and Chambers's *Information for the People* (1835) provide obvious examples. The accounts of the calendar offered here are scrupulously fair and surprisingly open in showing the significance of pagan traditions of ritual, the seasonal demands of the agricultural year, and the complexities of astronomical movements, rather than the shaping festivals of the Christian devotional year, in formulating the calendar. Indeed, within these accounts, the early Victorian year becomes a shape constructed out of pagan festivals, commercial need and Christian belief, all superimposed on the astronomical division of time. Lardner, whose *Museum* takes the cult of information to excess, devotes sixty-odd pages to explaining how the Christian year derives from astronomical movements rather than from biblical history and Christian liturgy.

The central presence of the SPCK, the SDUK, and the entrepreneurs of artisan self-education in these discussions is clear acknowledgement of their significance. Yet what comes across from these theorized histories of the calendar is an unexpected stress not on a year determined by Christian ritual but rather on a year which has been *culturally* determined through history, and thus become open to conflict and contest. If the aim of 'useful' almanacs was primarily to promote a regulated, rational, productive and anti-superstitious approach to the year among artisans, this apparently conservative purpose was only achieved by publicizing the calendar as a socio-historical construct with a divided and highly contested history. Such debates about the nature of 'Common Things' (to use the title of a volume of *The Instructor*), which characterize the literature of useful knowledge in the

THE
BRITISH ALMANAC,

OF

THE SOCIETY FOR THE DIFFUSION OF USEFUL
KNOWLEDGE,

FOR THE YEAR OF OUR LORD

1848,

BEING BISSEXTILE, OR LEAP-YEAR;

CONTAINING

THE CALENDAR OF REMARKABLE DAYS AND TERMS;

MONTHLY NOTICES, SUNDAY LESSONS;

METEOROLOGICAL TABLES AND REMARKS;

Astronomical Facts and Phenomena;

TABLES OF THE

SUN, MOON, AND TIDES;

WITH A

MISCELLANEOUS REGISTER OF INFORMATION

CONNECTED WITH GOVERNMENT, LEGISLATION, COMMERCE,
AND EDUCATION; AND

VARIOUS USEFUL TABLES.

LONDON:

CHARLES KNIGHT, 22, LUDGATE STREET.

SOLD BY ALL BOOKSELLERS IN THE UNITED KINGDOM.

Price One Shilling stitched,
Or Four Shillings bound in cloth with the 'Companion to the Almanac.'

Printed by A. SWEETING, Bartlett's Buildings, Holborn.

Figure 3: The title-page of *The British Almanac* for 1848

1830s and 1840s, seem to me to suggest how deeply controversial 'information' and 'fact' must be within any society.

5.

Overlaying the continuing production of these apparently utilitarian and informative almanacs aimed at artisans, traders, farmers and working people, a more self-consciously *literary* and *discursive* tradition of year-making began to emerge in the 1830s and 1840s, deriving from the popular success of William Hone's *Everyday Book* (1826–7) and *Table Book* (1827–8). The bibliography of these two texts is extremely complex – suffice it to say that they were extensively reprinted by Thomas Tegg and others, and, together with Hone's *The Year-Book of Daily Recreation and Information Concerning Remarkable Men and Manners, Times and Seasons* (1832), form an extraordinarily detailed, discursive and polemical account of the calendar. This sense of discursiveness is central to Hone's purpose. This is the declaration he makes in the introduction to *The Table Book*: 'In my *Table Book*, which I hope will never be out of "season," I take the liberty to "annihilate both time and space," to the extent of a few lines or days, and lease, and talk, when and where I can, according to my humour.'[21] In other words, Hone is using the almanac structure as an anecdotal occasion, and he builds all his year-book compilations out of antiquarian speculations and curiosities, accounts of ancient customs and beliefs, and other 'popular amusements' – what the sub-title of *The Everyday Book* calls 'sports, pastimes, ceremonies, manners, customs, and events'.

Two other aspects of Hone's work are also distinctively new. The first is Hone's attempt to include, indeed specifically to acknowledge, artisan readers as the proper audience for his nationalistic popular antiquarianism and thus to create a particular kind of English reading community. The second is his brilliant appropriation of the wood-engraved vignette for this purpose – the wood-engraved vignette was a form which alluded at the same time to traditions of broadside song, popular prose narrative and emblematic caricature, and in George Cruikshank, Hone found an artist capable of sustaining this complex of sophisticated and apparently vernacular possibilities. Hone, building on the kind of readership first defined by Paine, belong to a Tory/Radical tradition which also includes Cobbett and, later, Ruskin. His calendar books are sustained by the notion of a secular working year dominated by agriculture, but punctuated and lightened by festivals, fairs, commemorations and rituals, very few of which derived

THE

INSTRUCTOR:

VOL. IV.

CONTAINING

LESSONS ON THE CAL[ENDAR]
THE MONTHS.
AND THE SEASONS.

PUBLISHED UNDER THE DIRECTION OF
THE COMMITTEE OF GENERAL LITERATURE AND EDUCATION,
APPOINTED BY THE SOCIETY FOR PROMOTING
CHRISTIAN KNOWLEDGE.

PRICE TWO SHILLINGS

LONDON:
JOHN W. PARKER, WEST STRAND.

THE CALENDAR

AND

ALMANACK;

IDUNA.

THOR.

WODEN.

THE SAXON IDOLS, AFTER WHICH THE DAYS OF THE WEEK WERE NAMED.

LESSON I. *The Calendar and Almanack —Origin of the Divisions of Time.*

By the word *Calendar*, we understand *an account of dates, or a register of time.* An account of this kind includes all such particulars as relate to the different periods into which time has been divided, and by which it is reckoned; such as years, months, weeks, days, and hours. It is also usual to mark in the Calendar any remarkable event which may have happened on a particular day, or any thing connected with the seasons, or the appearances of nature.

The word itself, like many others which we use when speaking of time, is borrowed from the Romans. The first days of their months were called *Calends*, and the public notices or registers of coming dates and events, were called *Festi Calendares*. These were hung up in places where

[9]—2

Figure 4: Title-page and opening spread from *The Instructor*

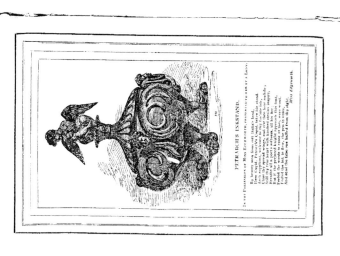

Figure 5: Title-page spread from *The Every-Day Book* (1830)

from or were related to the official calendars of Church and State. The seasons for Hone, as his introductory comments (quoted above) suggest, are not so much divisions of the agricultural year or reminders of devotional responsibilities as occasions for discursive pleasure, often expressed as public, if unofficial, carnival. Cruikshank's engravings, produced exactly at the moment when mass-circulation illustration was becoming a dominant form, mediate precisely between the vernacular pleasures of broadside and chapbook and the sophisticated iconography of caricature. They are both demotic *and* genteel, polite *and* unruly.

Hone's year, then, emerges as an agricultural calendar rewritten by a combination of Blake and Smollett. It constructs an Albion of festivals, customs and remembrances, which is independent both of conventional history and of the obligations, reminders and orthodoxies of the 'useful' almanac. This kind of Tory/Radical re-working of the year was picked up, if in rather muted and conciliatory forms, by a string of subsequent publications which sought to sustain notions of the British yeoman and peasant, the agricultural year, and sturdy, independent self-reliance learnt from Hone as an aspect of a celebratory agricultural year. Several of these were produced by William Howitt as year books, seemingly for artisans, but with sentimental Birket Foster illustrations and a generally diluted sense of Hone's sturdy defence of artisan culture.[22] The swan-song of the discursive artisan year book was Robert Chambers's enormously popular *Book of Days*, first published between 1862 and 1864.[23] These publications all allude to Hone, not just in their form but also in their defence of the possibility of alternative constructions of the social meaning of time built out of the stories, memories, daily tasks and lived experience of rural workers, rather than out of the compelling logic of clock time, work discipline, and the defiance of the hour of the day and of season implicit in factory production and the shift system. The notion of time predicated in these and similar texts needs more study, but I want to conclude this essay by looking at the ways in which the previously controversial genre of the almanac becomes assimilated into Victorian literary culture without major conflict.

6.

The key figure in these transitions is Hone's original illustrator George Cruikshank. Cruikshank's *Comic Almanac* was published each year between 1835 and 1853.[24] It forms a sequence of volumes which has

THE COMIC ALMANACK,

FOR 1840:

AN EPHEMERIS IN JEST AND EARNEST,

CONTAINING

"ALL THINGS FITTING FOR SUCH A WORK."

BY RIGDUM FUNNIDOS, GENT.

ADORNED WITH A DOZEN OF "RIGHTE MERRIE" CUTS,
PERTAINING TO THE MONTHS, AND
AN HIEROGLYPHIC,

BY GEORGE CRUIKSHANK.

LONDON:

IMPRINTED FOR CHARLES TILT, BIBLIOPOLIST,
IN FLEET STREET.

WHITEHEAD AND CO. PRINTERS, 76, FLEET STREET.

Figure 6: Title-page from *The Comic Almanack* for 1840

attracted enormous enthusiasm from Cruikshank's critics. R. A. Vogler concludes his account of 'this remarkable periodical' by hoping that a 'scholarly study will be made one day to accompany a complete reissue of this impressive work'.[25] He points out that Cruikshank made 196 single-plate etchings for *The Comic Almanac* as well as countless smaller illustrations. Vogler's main argument, which is certainly borne out by my own response, is that 'these small etchings epitomize Cruikshank's transformation of the Regency caricature into the format of Victorian book illustrations'.[26] R. L. Patten's recent exhaustive biography of Cruikshank included fifteen pages of commentary on *The Comic Almanac* which, like Vogler's account, stress both the inventiveness of the early volumes and the loss of energy in the later ones, a decline which contributed to Cruikshank's loss of his audience in his last working years.[27]

The Comic Almanac derives more from traditions of satirical image-making than from any acknowledgement of the profound and contested history of the almanac. Indeed, within Cruikshank's serial work (which did however contain many visual elements of the traditional almanac form), the almanac idea is used largely as an organizational convenience rather than as a significant way of apprehending the world. In this, Cruikshank is extending the development of the single-plate caricature tradition which, in the 1830s, is beginning to replace large, complex, single images with plates built up from multiple small images loosely organized as jokes on a single theme such as 'Seasons' or 'Fears', or visual puns based on nautical terms, to cite examples all drawn from the work of Henry Heath in the 1830s.[28] Thus in Cruikshank's hands the almanac idea becomes brilliantly safe, an opportunity for social satire, grotesque comic invention and visual pleasure, but not a site of ideological dispute or conceptual difficulty. Cruikshank's *Comic Almanac* is a firmly grasped opportunity for the artist to meet his relatively settled Victorian audience with something closer to comic illustration than social invective.

The sequence of *Punch Almanacs* which began in 1842 shows a similar opportunism by re-working the almanac as a diverting rather than a useful genre. The 1842 *Almanac* is an almost direct copy of Cruikshank's publication, but the apparent deference to the traditional almanac lay out and content proves on closer inspection to be merely facetious – '22 Feb. Tues BARRY died 1806. "Life let go his *painter*"'.[29] Funny, but hardly a piece of useful information or even a reminder of a necessary and important act of commemoration. But even these vestiges of the old almanac and its uses are quickly dropped. By 1849

Figure 7: A page from the *Punch Almanac* for 1842

the *Almanac* comprised a small annotated list of dates above a sequence of satirical 'years' – March and April offer the highly facetious 'Swell Mobsman's Almanack' – surrounded by visual jokes which are derived from contemporary social events rather than from any traditional seasonal significance. As topical satire this is effective. As

Figure 8: The opening page of Chambers's *Book of Days* [1862]

traditional almanac it is useless. By 1855 all that *Punch* offers of the old almanac is a small calendar set in a decorative emblematic border which summarizes popular festivals like St Valentine's Day and Christmas. The rest of the *Almanac* for that year looks like the normal weekly issues of the magazine.

I am not trying to demonize the early Victorian comic almanac as the means by which an older, more significant tradition of year-book compilation was overthrown or at least forgotten. Indeed, I am extremely interested in Cruikshank's *Comic Almanac* as a symptom of major changes in popular image-making and mass taste in the first years of Victoria's reign. It is important, too, to recall that the preceding account of the history of the almanac form is not made up of a simple sequence of developments, but rather one of superimpositions, negotiations, and overlappings. You could be reading *The Reformer's Almanac* in 1848, but you could equally well be reading the SDUK *Almanac*, or Cruikshank's *Comic Almanac*, or *Punch's Almanac*, or *The Pictorial Times Almanac*, or indeed any one of hundreds of others not represented here – or, probably, in any library in Britain. A copy of *Chambers's Book of Days* was to be found in many later Victorian households. By 1853, drawing on the list of one publisher only, you might have available *The Uncle Tom's Cabin Almanack; or, The Abolitionist Memento*; *The Illustrated Exhibitor Almanac*; *The Protestant Dissenters' Almanack*; *The Popular Educator Almanac*; or *The Temperance Almanack*.[30] I do think, however, that *Punch*'s retreat from the almanac's traditional function as an informative, predictive, sometimes discursive, but always utilitarian genre suggests that some of the social and political tension had gone out of the almanac and year-book genres by the 1850s. While the list of Cassell's almanacs above acknowledges and seeks to reiterate social attitudes and beliefs, there is also a discursive and decorative element in these publications which is more accommodating and inclusive than the polemics of, for example, *The Reformers' Almanac*. But tracing the history of the late-Victorian almanac – a history which would have to include *The Band of Hope Almanac* as well as John Hartley's dialect *Clock Almanac*, along with innumerable others – would need another essay altogether.

Notes

1. *The Annual Register for 1789* (London: G. G. J. and J. Robinson, 1790), p. 15.
2. Victor E. Neuburg, *Popular Literature: a History and Guide* (London. Woburn Press, 1977), p. 49.
3. See, for example, Thomas Gretton, *Murders and Moralities: English Catchpenny Prints, 1800–1860* (London: Colonnade Books, 1980), pp. 100–6. An annoying aspect of Gretton's otherwise fascinating book is that he prints all his chosen images without their surrounding letterpress text.
4. Louis James, *Print and the People, 1819–1851* (London: Allen Lane, 1976), pp. 49–61.
5. Keith Thomas, *Religion and the Decline of Magic* (London: Weidenfeld and Nicolson, 1971).
6. David Vincent, *Literacy and Popular Culture: England, 1750–1900* (Cambridge: Cambridge University Press, 1989), pp. 192–3.
7. Bernard Capp, *Astrology and the Popular Press: English Almanacs, 1500–1800* (London: Faber and Faber, 1979).
8. Lise Andries, 'Almanacs: Revolutionizing a Traditional Genre', in *Revolution in Print: the Press in France, 1775–1800*, eds R. Darnton and D. Roche (Berkeley: University of California Press, 1989), pp. 202–22.
9. Owen used almanacs in his model factories at New Lanark both as a means of informing his workers and of maintaining discipline and self-consciousness about conduct in the work place. See illustration 2; *The Reformers' Almanac* was published by its author Joseph Barker. It was accompanied by an even more polemical *Companion*. Kathleen Tillotson drew this publication to my attention, and I am grateful to her for the loan of a copy.
10. Ronald Hutton, *The Stations of the Sun: a History of the Ritual Year in Britain* (Oxford: Oxford University Press, 1996).
11. Maureen Perkins, *Visions of the Future: Almanacs, Time and Cultural Change* (Oxford: Clarendon Press, 1996). As Perkins acknowledges, behind all these recent discussions lies E. P. Thompson's famous discussion of 'Time, Work-discipline and Industrial Capitalism' first published in *Past and Present* in 1967.
12. Dionysius Lardner, *Museum of Science and Art*, 12 vols (London: Walton and Maberly, 1854–6). Volume VII is called *Common Things* (1855) and contains an extensive account of the history of the almanac; *The Instructor* was first published in the 1830s and much reprinted during the following two decades. The volumes included 'instruction' in basic literacy, history, natural history and geography, and are all illustrated by wood engravings. Volume IV comprises 'Lessons on the Calendar, the Months, and the Seasons'.
13. For the very little that is known about Grant see: Louis James, *Fiction for the Working Man* (Oxford: Oxford University Press, 1963), pp. 21, 53, and William Feaver, *Masters of Caricature* (London: Weidenfeld and Nicholson, 1981), pp. 11–12.
14. Perkins, *Visions of the Future*, pp. 46–88.
15. This illustration appears to be one of Thomas Onwhyn's series of extra illustrations to *Pickwick*. This series was published by E. Grattan in 1837. I am grateful to Ian and Felicia Gordon for finding and making available this image.

16. Turner was only the most famous of a long line of artists who were commissioned to draw for *The Oxford Almanac*. For a detailed account of Turner's contributions see Luke Hermann, *Turner Prints: the Engraved Work of J. M. W. Turner* (Oxford: Phaidon, 1990), pp. 14–17. Hermann's account serves as a useful introduction to the considerable significance of *The Oxford Almanac* in the history of topographical printmaking.
17. I am grateful to the Museum of Labour History, Manchester, and especially Dr Eddie Cass, for making a copy of this image available to me.
18. The image was actually drawn in 1834, and the date altered annually to maintain the commercial potential of the plate.
19. For some accounts of the role of these individuals and institutions committed to supporting the cultural development of working men and women see, among many other studies: R. K. Webb, *The British Working Class Reader* (London: Allen and Unwin, 1955); A. Lee, *Laurels and Rosemary: the Life of William and Mary Howitt* (Oxford: Oxford University Press, 1955); R. D. Altick, *The English Common Reader* (Chicago: University of Chicago Press, 1957); and D. Vincent, *Literacy and Popular Culture* (Cambridge: Cambridge University Press, 1989).
20. Dionysius Lardner, *Museum of Science and Art*, VII, 2.
21. William Hone, *The Table Book*, 2 vols (London: Hunt and Clarke, 1827), I, 2.
22. See, for example, Howitt's *The Book of the Seasons; or, Calendar of Nature* first published by Henry Colburn in 1830 and reprinted on several occasions, and *The Year-Book of the Country* (London: Henry Colburn, [n.d.]).
23. Robert Chambers, *The Book of Days*, 2 vols (Edinburgh: W. and R. Chambers, 1862–4). See illustration 8.
24. G. Cruikshank, *The Comic Almanac* (London: Charles Tilt and others, 1835–1853). The run of *The Comic Almanac* was republished by John Camden Hotten in two volumes, n.d.
25. R. A. Vogler, *The Graphic Works of George Cruikshank* (New York: Dover Books, 1979), p. 152.
26. *Ibid.*
27. R. L. Patten, *George Cruikshank's Life, Times and Art, 1835–1878*, 2 vols (Cambridge: Lutterworth Press, 1996), II, 189–203.
28. Henry Heath and his publisher Charles Tilt gathered together his miscellaneous engravings into a single volume called *The Caricaturist's Sketchbook* (London: Tilt, 1840).
29. *Punch*, 2 (1842), 2.
30. I have drawn this list from an advertisement in Cassell's *The Working Man's Friend*, n. s. 3 (13 November 1852), 111.

8
Tennyson and the Apostles
John Coyle and Richard Cronin

All of Tennyson's critics have recognized the importance to his poetic career of the years that he spent at Cambridge and of the friends that he made there. When he arrived in Cambridge on 9 November 1827, he emerged, as Herbert Tucker puts it, from 'the simultaneously empowering and stultifying atmosphere of his black-blooded father's rectory at Somersby into fellowship with the greater world'.[1] The Cambridge years were crucial in Tennyson's transformation from the poet described by Dwight Culler in the third chapter of his study of Tennyson as 'the solitary singer' into the poet Culler refers to in his fourth chapter, 'O Civic Muse'.[2] But the major poems do not so much provide the evidence that Tennyson has accomplished this transition as rehearse again and again its difficulty; from absorption in isolating grief for a dead friend to the expression on the occasion of his sister's wedding of a wholly communal joy, from the self-enclosed bitterness with which the speaker of *Maud* (1855) begins his poem to the chillingly complete subordination of personal to patriotic emotion with which it ends.

From very early in his career, it was the point of transition that held Tennyson's poetic interest; the moment when the Soul in 'The Palace of Art' (1832) feels herself 'shut out' from the world and begins to 'howl' for very loneliness, when the Lady of Shalott turns away from the mirror to look down on Camelot, or when Oenone determines to leave her enclosed valley for Troy to wreak her revenge on the man who has betrayed her. As a poet Tennyson prefers to take his stand not in an enclosed and private world nor in a public arena, but in the shadowy borderland between them. He is not a dreamer, and not a wide-awake poet either, but a poet fascinated by the process of waking, as in 'The Lover's Tale' (1879), when an apprehension of the external world 'Entering all the avenues of sense / Past through into his citadel, the brain, / With hated warmth of apprehensiveness' (ll. 619–21),[3] and fascinated too by the process of going to sleep, as in 'A Dream of Fair Women' (1832) when

> All those sharp fancies, by down-lapsing thought
>   Streamed onward, lost their edges, and did creep
> Rolled on each other, rounded, smoothed, and
>   brought
> Into the gulfs of sleep.
>
>   ll. 49–52

It was the in-between state that was to remain throughout Tennyson's career his preferred poetic territory, but it was in his years at Cambridge that he first claimed it.

From time to time throughout his life Tennyson joined clubs – the Sterling Club, the Cosmopolitan, the Metaphysical Society, the Athenaeum – and on each occasion he followed the same routine. He attended a few meetings and then allowed his membership to lapse. It was a pattern of behaviour that had begun early. At Cambridge, Hallam and all his most intimate friends were Apostles, members of the club formally known as the Conversazione Society.[4] Tennyson himself was formally admitted as an Apostle on 31 October 1829, and resigned less than four months later, after failing to fulfil his obligation to read a paper to the society (his was to have been on ghosts). From the date of his resignation, 13 February 1830, he occupied the ambivalent position that he seems always to have preferred, neither quite an Apostle, nor quite excluded from their circle, a benign version of the posture taken by his own Enoch Arden when he peeps through the window at the 'ruddy square of comfortable light' ('Enoch Arden' (1864), l. 722) that frames his wife, his daughter, her new husband and new baby. But for Tennyson, one suspects, the glow was the more comfortable for the pane of glass that separated him from it. It is the posture that informs many of his poems.

In Tennyson's day the Apostles liked to think of themselves as united not by a set of beliefs but by their common possession of 'the Apostolic spirit', a quality that enabled its members to disagree with one another on the most important matters without disturbing a good fellowship which was 'rather of brothers than of friends'.[5] In fact, the society that Tennyson joined was dominated by the influence of two members, F. D. Maurice and John Sterling, who had by 1829 left Cambridge for London where they were editing the *Athenaeum*,[6] and it seems clear that the younger Apostles still at Cambridge gave their broad support to the views that Maurice and Sterling promulgated in their journal. These can be briefly summarized. In religious matters the dominant influence was Coleridge. The Apostles recognized themselves as the nucleus of the 'clerisy' that Coleridge hoped might guide the nation's moral life. In philosophy, as one might expect, they were fiercely anti-Utilitarian. In poetry their allegiance was to Wordsworth and to Shelley, and in politics they were distinctly radical. Within a year Sterling and several of the younger Apostles, including Tennyson, were to involve themselves actively in the cause of the Spanish *Liberales*.[7] Tennyson's first opportunity to signal the impact on him of all this in a substantial poem came when he decided to compete for the Cambridge poetry prize.

In December 1828 it was announced that the subject of the poem for the Cambridge prize poem that year would be 'Timbuctoo'. It was a topical choice. The announcement came just a week after the French adventurer, René Caillie, had successfully claimed the reward that the French Geographical Society had offered to the first traveller to return from Timbuctoo, but the topic must surely have been decided by the Cambridge authorities before his success was known. For thirty years Timbuctoo had been the goal of a long series of expeditions. In the past decade an expedition led by Hugh Clapperton had failed, and Clapperton himself had died of disease before his party could extricate itself from the interior of Africa. In 1826, another Englishman, Major Laing, had actually reached Timbuctoo, but was murdered three days after leaving the city. The Cambridge authorities were offering their undergraduates the opportunity to write a poem patriotically celebrating the achievements of these men.[8] Both Tennyson and Arthur Hallam chose to compete for the prize, Hallam by submitting a poem that impudently challenged the expectations of his examiners.[9] In Hallam's 'Timbuctoo' the celebration of the daring of men such as Clapperton and Laing is always tempered by a recognition that, whether they knew it or not, these men were scouts trail-blazing for empire. England and France were not, Hallam knows, engaged in a quixotic race to be the first to discover one African city, but in a competition for the division of Africa. Timbuctoo makes its final appearance in Hallam's poem invested with the pathos of a land as yet innocent of European vices – the urge to dominate, the greed for gold – which are about to destroy it. Invited to join in a patriotic celebration of English heroism, Hallam responded by offering a Shelleyan condemnation of empire, a brave decision, and one which may have cost him the prize. Tennyson's response was quite different, but characteristic. He sent home for the manuscript of 'Armageddon', a fragment but the most substantial product of the years spent in the 'empowering and stultifying atmosphere' of the Somersby rectory. 'Armageddon' begins and ends with a theatrically apocalyptic landscape, but this acts only as a frame for the central event. A seraph alights beside the poet, and the poet gazes into the seraph's eyes. It is a scene of annunciation in which the poet is made aware of his own high vocation, and learns to recognize his own power. But it is a power that is fully expressed simply by virtue of being recognized:

> Yea! in that hour I could have fallen down
> Before my own strong self and worshipped it.
>
> Tennyson, 'Armageddon', ll. 49–50

This is a talent that is committed only to its own display, and it could not easily be otherwise, for Tennyson, as a young man at Somersby, already recognized his gift, but as yet could do nothing other than proclaim it, for he did not have anything to say. Tennyson found in his Apostolic friends at Cambridge not only the appreciative audience that every young poet needs, but a loosely consistent set of values that his poems could articulate. He was in need of the second quite as much as the first, and it might have been expected that when he quarried 'Armageddon' for the materials to make up his own entry for the competition he would, like Hallam, have placed his poem at the service of some acceptable Apostolic moral. Instead, Timbuctoo is offered by Tennyson as the last in the line of those ideal places that have haunted the imagination of poets, from the Hesperides to Xanadu, but it has a special potency for him because the city of the imagination is about to be superseded by the paltry settlement reached by Caillie; the city through which a river winds 'imaging / The soft inversion of her pendulous domes' is about to

> Darken and shrink and shiver into huts,
> Black specks amid a dreary waste of sand,
> Low-built, mud-walled, Barbarian settlements.
>
> 'Timbuctoo', ll. 242–4

The poem ends with a painful and exhilarating shock as Tennyson is wrenched from poetic dreams to a waking acknowledgement of a hard and meagre reality in a version of the plot that will underlie so many of his future poems. Tennyson does not write a poem that encapsulates the new ideas that he had found in the 'greater world' of Cambridge, so much as a poem that dramatizes the process of the emergence from the dreamy enclosed world of his Lincolnshire boyhood.

Tennyson's Apostolic friends had a very clear idea of the kind of poetry they wanted him to write; public, morally earnest, and trenchant. One of them, J. W. Blakesley, wrote to encourage him in his task:

> The present race of monstrous opinions and feelings which pervade the age require the arm of a strong Iconoclast. A volume of poetry written in a proper spirit, a spirit like that which a vigorous mind indues by the study of Wordsworth and Shelley, would be, at the present juncture, the greatest benefit the world could receive.[10]

Tennyson's failure to oblige sometimes provoked indignation: 'Why will not Tennyson give up absurdities of every kind – the errors of his own morbid, Germanized, and smoke-sodden temperament; and set about writing like a man?'[11] But Tennyson, one suspects, resigned his formal membership of the Apostles the better to maintain his jealous guard of the absurdities and temperamental errors that he knew to be integral to his talent.

Tennyson, on just one occasion, did agree to write a committedly Apostolic poem, the poem published in *Poems, Chiefly Lyrical* (1830) under the title 'A Character'. The target of this satiric squib is Thomas Sunderland, who had become an Apostle in December 1826, but by 1828 had become the society's preferred hate object, referred to as a Jonah, a Judas, the Beast of the Apocalypse. It is no longer possible to discover quite what he had done to cause such offence. It may simply have been that he disliked the habit of mutual flattery that the Apostles had developed to secure their brotherhood. He certainly complained to Monckton Milnes of the 'vulgarly gregarious' nature of the other members.[12] In Tennyson's poem he is represented as a man who contrives, by assuming attitudes and opinions only for effect, to offer an unseemly parody of the true Apostolic spirit. But Tennyson's poem seems not quite to sustain its satiric contempt:

> Most delicately hour by hour
> He canvassed human mysteries,
> And trod on silk, as if the winds
> Blew his own praises in his eyes,
> And stood aloof from other minds
> In impotence of fancied power.
>
> 'A Character', ll.19-24

The poem seems itself to share the aloofness that it castigates in its subject. It is an effect that Leigh Hunt, who thought the poem 'a faultless composition' of its kind, drew attention to when he pointed out how 'the delicate *blank* effect of the disposition of the rhymes completes the seemingly passionless exposure of its passionless object'.[13] It is, after all, a very similar aloofness that Tennyson was later to ascribe, with much more evident ambivalence, to the Soul in 'The Palace of Art'.

Tennyson found at Cambridge and amongst his Apostolic friends the set of opinions that he needed to give his poems a content, and that he needed, too, if his poems were to address their readers, but he seems to have known from the first that he both needed those opinions and needed to retain a certain detachment from them. The best evidence of this is in Tennyson's relationship to the most local of the campaigns that Maurice and Sterling conducted in 1828, for the reform of the Universities.[14] Maurice and Sterling attacked the standards of scholarship and of teaching that obtained at Oxford and Cambridge, offering the German universities as counter-examples. They derided the meagre demands made of those who took a pass degree, the work for which could be completed in five or six weeks of cramming immediately before the exams. They called attention to the narrowness of even the Honours syllabus, and the absurdity of excusing noblemen from any examination at all: 'it probably being supposed with regard to them that the business of legislation and the management of a large estate require less extensive information, and less matured faculties, than the affairs of meaner men'.[15] But the central charge was that the two universities were both socially and religiously exclusive, socially in that Oxford and Cambridge were universities 'devoted exclusively to the more powerful classes of society',[16] and religiously because attendance at chapel was compulsory, and because no man could matriculate at the University of Oxford without subscribing to the thirty-nine articles or graduate from Cambridge without vouching that he was a bona fide member of the Church of England. Maurice's and Sterling's campaign was cued by the founding of the University of London, which in the autumn of 1828 admitted its first students, and for the first time made an English university education available to non-conformists. An educational institution, they argued, could claim to be a 'university' only if it satisfied the claims implicit in that title; only that is, if it were a place for 'the communication of every kind of knowledge', and 'open to every class of people'.[17] By this definition it was clear that the University of London was not the third English university, but the first.

The most dramatic signal of support amongst Cambridge undergraduates for Maurice's and Sterling's position came in March of 1829, when the Cambridge Union debated the motion, 'It would have been expedient to have emancipated the Catholics in the year 1808' (the precise form of the motion was dictated by the University prohibition on debating political events of the past twenty years). After the repeal the previous year of the Test and Corporation Acts against Dissenters, Catholic emancipation, a measure that was to be carried through parliament by a reluctant Duke of Wellington in the following year, was an obviously topical issue. But it had too a more local significance, because, were the Catholics, like the Dissenters, to be granted full citizenship, the University of Cambridge would be in the embarrassing position of upholding the principle of religious restriction that had been abandoned by the parliament under which it held its charter. It was surely this that persuaded the Master of Trinity, Christopher Wordsworth, to be present at the debate and to congratulate those who spoke against the motion, and no less than four Apostles to speak in its favour. The motion was in the end carried by a small but decisive majority.[18]

Tennyson's most direct response to this campaign was the sonnet, 'Lines on Cambridge of 1830'. The poem savours the richly reclusive atmosphere of the University, its 'Wax-lighted chapels, and rich carven screens', with a Keatsian relish that quickly turns to mockery of 'Your doctors, and your proctors, and your deans'. The University has become an anachronism – 'your manner sorts / Not with this age wherefrom ye stand apart' – and will not survive the dawn of a new age, 'when the Day-beam sports / New-risen o'er awakened Albion'. The poem appears to end somewhat inconsequentially. The University will not survive,

> Because the lips of little children preach
> Against you, you that do profess to teach
> And teach us nothing, feeding not the heart.

But the lines are more pointed than they at first seem. They draw a distinction between teachers and those who merely 'profess' to teach, between those who speak to the heart sometimes through the lips of children and those who occupy themselves in seeking to ensure that a large proportion of those little children will not be suffered to enter the University of Cambridge. The implicit contrast is between Wordsworth, the poet rivalled only by Shelley as an object of Apostolic veneration, and his brother, Master of Trinity, Professor Christopher Wordsworth.[19]

This sonnet reveals more clearly than any other of Tennyson's poems how completely he had absorbed the values shared by his Apostolic friends: their moral earnestness, their pious distrust of established religion, their concern to address themselves directly to the age, and their political radicalism. But he chose not to publish it, either in the *Poems, Chiefly Lyrical* of 1830, or the *Poems* of 1832. In fact, it was not published until 1879 in R. H. Shepherd's unauthorized *Tennysoniana*. It was surely not a decision that marked a critical judgement, because both volumes include far weaker poems, nor does it seem likely that Tennyson was nervously unwilling to court controversy. It seems more likely that he reached the conclusion that a poem that so decisively rejected 'wax-lighted' spaces for the 'Day-beam' of the open air simply did not quite 'sort' with him.

The poem that he did publish, in *Poems* (1832), that seems linked to the campaign for the reform of the universities is far less direct. In 'The Palace of Art', as many critics have noted, the architecture of the palace is based on the Great Court of Trinity College, and the life that the Soul lives there, absorbed in a detached, aesthetic contemplation of the palace's contents, which themselves seem designed as a rather comprehensive museum of human culture, is a life recognizably akin to that of an idealized undergraduate. The unpeopled rooms through which the Soul wanders are very like the 'vacant courts' in the sonnet on Cambridge, and it is the Soul's decision just as it is the University's to 'stand apart' from the age. The period in which the Soul resides in her palace even corresponds precisely to the common length of an undergraduate career. She leaves only when 'four years were wholly finished', discarding her 'royal robes' as she leaves just as an undergraduate might throw away his gown. There is perhaps even a vestige of Maurice's and Sterling's attack on the exclusiveness of the University in the poem's final lines, when the Soul entertains the notion that she might at some point return to the palace, but that if she does so she will return 'with others'. Tennyson has taken the public campaign that

Maurice and Sterling had mounted, and taken, too, the public values that had impelled it, but in 'The Palace of Art' such things are refracted within a narrative, that, like so many of Tennyson's narratives, works to frame its most distinctively Tennysonian moment, the moment when the Soul is suspended between the rich stagnancy of her self-absorption and all the 'moving Circumstance' of the world outside, like the 'still salt pool' abandoned by the tide

> that hears all night
> The plunging seas draw backward from the land
> Their moon-led waters white.
>
> 'The Palace of Art', ll. 250–2

In these early Cambridge poems Tennyson works out his own solution to the problem of how to place the jealously guarded private world in which he knew his talent had its source in some relation to the public world around him, the problem of how he might best find a way to 'sort with his age'. It was no doubt for Tennyson a temperamental problem that had its origins, as Tucker suggests, in 'the simultaneously empowering and stultifying atmosphere of his black-blooded father's rectory at Somersby', but it was also a much more general problem to be confronted by all Victorian poets who accepted that a poem was authentic only in so far as it maintained a lyric voice, and yet aspired to write a poetry that addressed the circumstances of their times. The solution that Tennyson found, and that served him throughout his career, was to avoid any direct linkage between the private and public worlds, but instead to move between them by what he calls in 'The Princess' (1847) 'a strange diagonal' ('Conclusion', l. 27). It was not a solution that immediately satisfied his Apostolic friends, who repeatedly urged him to cultivate a poetry of more direct statement, and yet if Tennyson had not attended to their admonitions, and absorbed something of their high-minded public-spiritedness, he might never have sought a solution at all, and have remained throughout his life a minor lyric poet, a more talented version of his brother, Charles.

In this, too, 'The Palace of Art' is an exemplary poem, suggested as it was by a remark made to Tennyson by an older Apostolic friend, Richard Chenevix Trench, who once reproved him by remarking, 'Tennyson, we cannot live in Art'.[20] But his relationship with his friends was not one-sided. If he listened to them, they listened to his poems. In a letter of November of 1830, Blakesley rather charmingly records his difficulty in learning how to 'hear' Tennyson's rhythms: 'The worst of Tennyson's poetry is that it is necessary to hear him read it before you can perceive the melody, at least for ninety-nine men out of a hundred'.[21] But clearly he worked hard to adjust his ears to the new music, and, as with the rhythm of the poems, so it was with their content. Tennyson's Cambridge friends learned from studying the poems sufficiently to moderate their demand that the poet should act as the mouthpiece for approved moral principles. Even Blakesley, who had once required Tennyson to equip himself with 'the arm of a strong Iconoclast', began to suspect that a poem might have too much of a moral. By the time that he read Trench's 1834 volume, *The Story of Justin Martyr and Other Poems*, he was at least as aware of a contrary danger, that the poet might be 'swallowed up' in the clergyman.[22]

It was at Cambridge that Tennyson first learned that a poet need not choose between ignoring his readers or surrendering to their expectations, but that it remained possible to negotiate with them, and it was through these negotiations that Tennyson arrived at the paradoxical position that, by virtue of his celebrity, came to define for the Victorian age and beyond the peculiar nature of the poet's authority. Poets were granted a special authority to speak to and for the age, but, unlike the novelist, the authority was conditional on the poets' maintaining themselves at a remove both from their audience and from their times.

Notes

1. Herbert F. Tucker, *Tennyson and the Doom of Romanticism* (Cambridge, MA: Harvard University Press, 1988), p. 7.
2. A. Dwight Culler, *The Poetry of Tennyson* (New Haven: Yale University Press, 1977).
3. All quotations from the poems are taken from *The Poems of Tennyson,* ed. Christopher Ricks, 3 vols (London: Longman, 1987).
4. On the Cambridge Apostles, see Peter Allen, *The Cambridge Apostles: the Early Years* (Cambridge: Cambridge University Press, 1978).
5. Connop Thirwall quoted in Peter Allen, *The Cambridge Apostles: the Early Years*, p. 19.
6. They finally bought the journal in July 1828, and merged it with the *Literary Chronicle* that they already owned. But they had been the major contributors to the *Athenaeum* for some months before that.
7. For accounts of this affair and of Tennyson's involvement in it, see A. J. Sambrook, 'Cambridge Apostles at a Spanish Tragedy', *English Miscellany*, 16

(1965), 183–94; Robert Bernard Martin, *Tennyson: the Unquiet Heart* (Oxford: Clarendon Press, 1980), pp. 115–22; and Richard Cronin, 'Oenone and Apostolic Politics, 1830–2', *Victorian Poetry*, 30 (1992), 229–46. It marked a decisive and bloody end to the brief period of Apostolic radicalism, acting, as Carlyle memorably describes it in his *Life of John Sterling*, as 'the grand consummation and explosion of Radicalism in his life; whereby, all at once, Radicalism exhausted and ended itself, and appeared no more there' – *The Works of Thomas Carlyle*, 30 vols (London: Chapman and Hall, 1896–9), IV, 53. By 1832, Hallam was busy organizing a petition of Cambridge men against the Great Reform Bill. The change in Apostolic politics is most amusingly confessed by Richard Chenevix Trench in a letter to John Kemble of 16 July 1832, in which Trench reveals: 'I have given over despairing, and reading Shelley, and am beginning to acquiesce in things just as they are going on'. He adds: 'What are your reactions about the Reform Bill? I confess myself much alarmed, and do not look at it with that eye of favour which everybody seems to expect one should' – *Richard Chenevix Trench: Letters and Memorials*, ed. Mary Chenevix Trench, 2 vols (London: Kegan Paul, Trench, 1888), I, 97.
8. The topic for the poem was announced in *The Times* of 13 December. Entries were to be submitted by 31 March 1829. On the search for Timbuctoo, see Brian Gardner, *The Quest for Timbuctoo* (London: Cassell, 1968). The fate of Major Laing was revealed in the *Quarterly Review* in an account of Clapperton's expedition. See the *Quarterly Review*, 39 (1829), 143–83.
9. Hallam's poem is printed in *The Writings of Arthur Hallam*, ed. T. H. Vail Motter (London: Oxford University Press, 1943), pp. 37–44.
10. Hallam Tennyson, *Alfred Lord Tennyson: a Memoir*, 2 vols (London: Macmillan, 1897), I, 68. The letter is dated simply 1830.
11. Aubrey de Vere, quoted in James Pope-Hennessy, *Monckton Milnes: the Years of Promise, 1809–1851*, 2 vols (London: Constable, 1949), I, 118.
12. Sunderland's career is sketched by Peter Allen, *The Cambridge Apostles: the Early Years*, pp. 40–55. Allen quotes from the letter to Monckton Milnes, p. 51.
13. *Tennyson: the Critical Heritage*, ed. John D. Jump (London: Routledge and Kegan Paul, 1967), p. 131.
14. See the following articles, all in the *Athenaeum*, 1 (1828): 'Observations on Universities', 37 (9 July), 575–6; 'The University of Göttingen', 44 (27 August), 687–9; 'To the Bishop of London on the New Universities', 47 (17 September), 746–7; 'London University and King's College', 51 (15 October), 799–800; 'A Few More Words on the New Universities', 56 (19 November), 887–8; 'The New Universities', 57 (26 November), 905–6; 'Cambridge', 58 (3 December), 911–12; 'Cambridge, No. II', 60 (17 December), 943–4.
15. *Athenaeum*, 1 (1828), 911.
16. *Ibid.*, 911.
17. *Ibid.*, 799.
18. The debate is described by Peter Allen, *The Cambridge Apostles: the Early Years*, pp. 48–9.

19. The best commentary on Tennyson's sonnet is provided by a typically exuberant letter from John Kemble to his fellow-Apostle William Bodham Donne, quoted by Peter Allen, *The Cambridge Apostles: the Early Years*, p. 100. Kemble vows to lend his hand:

> to the great work of regenerating England, not by Political Institutions, not by extrinsic and conventional forms! By a higher and a holier work, by breathing into her the vigorous feeling of a Poet, and a Religious man, by pouring out the dull and stagnant blood which circulates in her veins, to replenish them with a youthful stream, fresh from the heart: yea so be it, even must the cost be my own lifeblood. This task not I alone must lay before me. To you, to Trench and Sterling, to Maurice, to all that band of surpassing men is the labour confided, and the commandment given.

He adds: 'Wordsworth has begun the work, he has delivered the sown field into our hand, and is not the harvest ours?', and concludes: 'we must strike through Education, and first at the Universities'.
20. Hallam Tennyson, *Lord Tennyson: a Memoir*, I, 118.
21. *Richard Chenevix Trench: Letters and Memorials*, I, 50.
22. Blakesley's reaction is recorded in James Pope-Hennessy, *Monckton Milnes: the Years of Promise, 1809–1851*, I, 14.

9
*Twist*ing the Newgate Tale: Dickens, Popular Culture and the Politics of Genre

Juliet John

> Mr. Long Ned, Mr. Paul Clifford, Mr. William Sykes, Mr. Fagin, Mr. John Sheppard, [...] and Mr. Richard Turpin [...] are gentlemen whom we must all admire. We could 'hug the rogues and love them,' and do – *in private*. In public, it is, however, quite wrong to avow such likings, and to be seen in such company.[1]

William Thackeray's coy confession of admiration for the criminal heroes of the so-called 'Newgate novels' is at the same time a witty sidesweep at Victorian mores. In the 1830s and 1840s, so-called 'Newgate' fiction and melodrama caused a welter of critical controversy by foregrounding criminals who were perceived to be attractive. The comparative lack of critical hostility to eighteenth-century crime fiction suggests that there were specific textual and contextual factors at stake in the Newgate controversy. Focusing on *Oliver Twist*, this essay will argue that the controversial cult status achieved by Newgate texts and protagonists had very little to do with the ethics of character in the novel, and very much to do with anxieties about power – in particular the power of 'popular' culture at the dawn of a 'modern' age.

But to begin with some simple definitions. The term 'Newgate', of course, refers to both the famous prison destroyed by fire in 1780, and to *The Newgate Calendar; or, The Malefactors' Bloody Register*, a popular collection of criminal biographies published in 1773. In his comprehensive work, *The Newgate Novel*, Keith Hollingsworth explains that as a literary critical term, the 'Newgate' tag is nothing but a convenient historical label. In practice, it was used insultingly by contemporary commentators about a series of novels published between 1830 and 1847 which had 'criminals as prominent characters'. According to Hollingsworth, 'a book was not likely to be damned with the accusing name unless it seemed to arouse an unfitting sympathy for the criminal'.² The texts which aroused such concern included, most famously, *Paul Clifford* (1830), *Eugene Aram* (1832) and *Lucretia* (1846) by Edward Bulwer, and *Rookwood* (1834) and *Jack Sheppard* (1839–40), by William Harrison Ainsworth.³ All were, in the words of John Sutherland, 'sensationally popular'.⁴ It is difficult now to imagine the cultural phenomenon that the Newgate novel represented; in 1840, the murderer B. F. Courvoisier is rumoured to have blamed his crime on his reading of *Jack Sheppard*, with the result that further stage adaptations of the novel were (unofficially) prohibited;⁵ and in 1852, juvenile delinquents told a House of Commons Select Committee on Criminal and Destitute Juveniles that Jack Sheppard had influenced their crimes.⁶

As *Oliver Twist* was published in serial form between 1837 and 1839 in *Bentley's Miscellany* – the same journal that published the 'Newgate' novels of Ainsworth – it was in some ways inevitable that it would be labelled a Newgate novel. Indeed, the fact that Dickens chose to write *Oliver Twist*, with its veritable rogues' gallery, despite critical antipathy to books about criminals, shows a typical Dickensian blend of courage and opportunism – that is, controversy sells. At first glance, however, it seems surprising that *Oliver Twist* – a novel whose protagonist represents, Dickens tells us, 'the principle of Good surviving through every adverse circumstance' (p. lxii)⁷ – should be tarred with the same critical brush as novels featuring 'heroes' like Paul Clifford or Dick Turpin, the daring highwaymen. Villains like Sikes, Fagin and Monks, however changed by subsequent musical metamorphoses, are not overtly glamorous. The closest Dickens gets to the seemingly innocent attractiveness of some of the Newgate heroes is the children Charley Bates and the Artful Dodger (to whom I shall return), who seem to play at the margins of the text.

So why did critics object to *Oliver Twist* – strongly enough for Dickens to feel the need to add a Preface to the 1841 edition of the novel vigorously defending the novel against charges of immorality? In the vanguard of the attacks on the Newgate novel was *Fraser's Magazine*, a radical, Tory, aggressively middle-class publication, which made an enemy of Bulwer in particular – partly because of the personal animosity he inspired in its editor, William Maginn, and major contributor, William Makepeace Thackeray. The objections to *Oliver Twist*, however, are out of step with other reviews of the novel, which were largely favourable when the first edition was published in book form.[8] In his contribution to a series of articles published in *Fraser's* protesting against the fictional romanticizing of crime, Thackeray groups Dickens with the Newgate novelists on the grounds that his portrayal of low-life villainy is caricatured and therefore unrealistic:

> We (that is, the middling classes) have been favoured of late with a great number of descriptions of our betters, and of the society which they keep; and have had also, from one or two popular authors, many facetious accounts of the ways of life of our inferiors. There is in some of these histories more fun – in all, more fancy and romance – than are ordinarily found in humble life; and we recommend the admirer of such scenes, if he would have an accurate notion of them, to obtain his knowledge at the fountain-head, and trust more to the people's description of themselves, than to Bulwer's ingenious inconsistencies, and Dickens's startling, pleasing, unnatural caricatures.[9]

The rationale behind Thackeray's criticisms of *Oliver Twist* seems to be that because Dickens's criminals are unrealistic (in style and conception), they are therefore immoral. Dickens's Preface to the third edition of *Oliver Twist* shows that he had taken such criticisms very much to heart; here he stresses the fidelity and 'truth' of his representation of low life, pitting his own writing against 'Romance' in a manner unusual for Dickens. We are after all talking about the author of *Hard Times* (1854), the writer whose style constantly evokes 'the romantic side of familiar things' (to quote the Preface to *Bleak House* (1852–3), p. xiv);[10] from the epigraph to *Sketches by Boz* (1836) to the late essay, 'The Spirit of Fiction',[11] Dickens demonstrates a sophisticated awareness of the relationship between reality and fiction, evincing indeed a (pre-) 'postmodernist' consciousness that reality itself can be a fictional construct.

The Preface is best understood as a piece of literary propaganda in a local critical debate. Newgate novelists and critics alike were grappling semi-consciously with the phantasmic concept that was to become 'realism' before the term was current in literary criticism, opposing it to the term 'romance'. On the most obvious level, the critical obsession with the relationship between the romantic and the real which surfaced in the 1830s and continued unabated for much of the period[12] is an aesthetic debate, an attempt to define and delineate the powerful new genre, the novel. The fact that controversy surrounded the Newgate novels at all is perhaps surprising when one considers that critics on all sides seemed to adhere to Thackeray's aesthetic rationale; put simply, that realistic fictional representations are 'good' while romantic representations are 'bad'. The obvious conclusion, therefore, is that while most writers at the time agreed that reality in the novel was a good thing, there was severe disagreement about the nature of reality itself. This is perhaps unsurprising considering the shockwaves that were still being felt from the French Revolution and the war with France (1793–1815), and the radical social and political changes which occurred in the 1830s: changes in the legal system which transformed Britain from a punitive to a 'disciplined' society, to use Foucauldian terminlogy, the Chartist movement, and of course the 1832 Reform Bill are just some of those that could be mentioned here.[13] 'Modernity', it could be argued, had begun.[14]

In 1837, the year that *Oliver Twist* appeared and Victoria came to the throne, the country was also in the midst of a cultural revolution quieter perhaps than the political and industrial revolutions afoot, but no less formative in the development of a 'modern' state. Literacy was increasing and developments in the publishing trade meant that books and newspapers were expanding their readership, moving further down the social ladder in the early Victorian period than ever before so that they 'began to reach the same class which the Reform Bill enfranchised'.[15] The availability of cheap fiction imprints, the dramatic proliferation in the number of cheaper literary journals in circulation, and the increasing tendency to serialize novels in periodical form (which surprisingly worked out cheaper for the 'consumer'),[16] meant that 'the democratization of politics was not only reported but reflected in the press'.[17] The popularity of penny dreadfuls, cheap weeklies and the practice of 'extracting' and serialization in newspapers all played their part in forcing down book prices, a trend which led to a broader readership, and which continued until 1850. Dickens's novels were, of course, formed by, and formative in, these crucial changes in the distribution of cultural capital. Though serialization in newspapers was never as prominent in Britain as in France, many of Dickens's *Sketches by Boz* were originally published (and almost simultaneously 'extracted') in newspapers (1834–5), and with his later novels, played a major role in the popularizing of fiction and newspapers.

These profound changes in communication systems and cultural ownership caused as many anxieties in the early Victorian period as the revolution in information technology has caused in our own. Indeed, in an eerie echo of now-common complaints about the power of the press, Edward Bulwer complains in 'A Word to the Public' (1847), 'The essential characteristic of this age and land is *publicity*'.[18] Victorians were encountering ideological and ethical problems which the modern mass media has made all too familiar; arguably, they were witnessing the beginnings of that very mass media. The story of the Newgate novel is inextricably entangled with the history of the mass media because the stories they related were 'sensationally popular' at a crucial juncture in the history of popular culture. It could no longer be assumed that popular culture was an uncomplicated expression of the tastes and values of the populace as the emergence of the machinery and capitalist economic dynamics of modern 'mass culture' was accelerating.

Bulwer's attack on the press on which he depended for his popularity suggests his own deep ambivalence about his popularity and, more importantly, the ambiguous literary status of the novel at a time when the Newgate controversy gave it such cultural centrality. The fact that the rise of the novel as a genre shadows the political and economic rise of the middle class has been securely established by Ian Watt, and more recently, Nancy Armstrong.[19] It is worth reminding ourselves, however, that neither the novel nor the middle class was as established when Dickens began *Oliver Twist* as blanket accounts of literary history sometimes lead us to believe. In the early 1830s, for example, reviews of biography, criticism and non-fiction prose (not to mention the usual suspects, poetry and drama) were all included in the 'Literature' colums of newspapers, while fiction was all included under the headings 'Magazine Day' or 'Miscellaneous'.[20] Indeed, Chittick makes the valuable point that the *The Pickwick Papers* (1836–7) and *Oliver Twist* were not originally conceived as novels, but as a 'periodical' in the case of the former, and a 'serial' in the case of the latter.[21]

The immense popularity of the Newgate novels and the burgeoning media which allowed them to reach a wider audience than ever before no doubt helps account for the revolutionary rise of the novel genre up the literary hierarchy. Popularity alone, however, would not be enough to alter the generic *status quo* in this way, as the history of the early-Victorian theatre makes clear. Stage melodrama, for example, was also hugely popular and undoubtedly increased the cultural impact of Newgate stories with its many adaptations of Newgate fiction for the literate and illiterate alike. The popularity of these plays did nothing to rescue the reputation of the theatre; it rather aroused anxieties ostensibly about the 'quality' and moral influence of these plays. In reality, however, the fact that melodramas in working-class and artisan theatres could not be properly policed was of major concern.[22] Though much theatre criticism emphasizes the conservatism of the melodrama which was so popular at this time, the very presence of the working classes in the theatres in unprecedented numbers to watch plays that were often overtly political was a cause of real anxiety for elements of the political and theatrical establishments. In the end, the 'legitimate' theatres survived by incorporating melodrama into their programmes. It is possible to see this U-turn on the part of the legitimate theatres as either a conservative act of cultural appropriation by the bourgeoisie or an instance of subversive upward mobility on the part of the illegitimate, melodrama. Indeed, if we consider the cultural history and dissemination of the Newgate biographies of criminals like Dick Turpin, Eugene Aram and Jack Sheppard, it is again possible to trace either a pattern of appropriation whereby oral narratives are eventually incorporated – via 'street literature' and the *Newgate Calendar* – into the bourgeois Victorian novel, or a process by which street culture forces itself, radically and subversively, into the Victorian drawing-room.

A genuinely popular art form which aims to educate the literati and layperson alike is obviously a formidable tool of power. Indeed, in many ways, the Newgate controversy is all about power – in particular, the power of fictional representations of reality and, perhaps more importantly, popular culture. This fact was not lost on the participants who had none of the coyness of the modern liberal in a television age about the 'power' of popular fictions. The Newgate controversy brought home the power of the novel with such force that each of the Newgate novelists was fully conscious of the power at their disposal – none more so than Dickens.

Oliver Twist is alone among the Newgate novels in analysing the role of the story-teller, entertainer or purveyor of fictions in the power dynamics of 1830s Britain. For *Oliver Twist* offers a sustained self-reflexive exploration of both Newgate fiction and the function of the entertainer in social structures of oppression. Crucial in this investigation is the relationship between 'reality' and fiction, and the role of each in the construction of the other. Contrary to the propagandist 1841 Preface, the 'realism' of the novel (in the sense of its photographic truth-to-life) is largely irrelevant to Dickens's ideological and moral scheme in *Oliver Twist*. The most sophisticated layer of commentary depends, not on its exact representation of life, but on its self-referential, textual investigation of the ideological and moral complexity of the relationship between life and fiction.

Fagin is central to this critique in so far as he possesses an acute understanding of this complexity and of the way fictions can be manipulated to achieve one's purposes – that is, to achieve power. Fagin's cynical deconstruction of life/fiction boundaries is compelling precisely because his construction as a character derives from a similarly sophisticated play with the same boundaries. Though based on a real-life prototype, Ikey Solomons, the character of Fagin obviously borrows heavily from the stereotype of the stage Jew[23] and the reader's response to Fagin relies on his/her recognition of, and openness to, well-worn theatrical, literary (and racist) conventions. Thus, to reapply Roland Barthes's term, the 'reality effect'[24] created by the character/caricature of Fagin is highly dependent on the reader's knowledge of the medium of fiction from the beginning.

It is through Fagin's relationship with the Artful Dodger and Charley Bates, above all, that Dickens answers the accusations that were to be levelled at him in the Newgate debate. The Artful Dodger and Charley Bates can be seen as fictional representations of the kind of boys investigated in the 1852 House of Commons inquiry into 'the situation of Criminal and Destitute Juveniles', who blamed their corruption on stage adaptations of Newgate novels. Bates and the Dodger are also the only characters in *Oliver Twist* who appear to be *overtly* attractive criminals in the same way as Robin Hood or the protagonists of the contemporary Newgate novels. They are thus crucial to Dickens in his textualized critique of Newgate fiction on two basic levels.

But to return to Fagin. Fagin is conscious from the outset that fiction, drama and comic entertainment have the power to corrupt. There are six instances that I wish to examine of Fagin playing a key role in Dickens's self-reflexive analysis of the Newgate controversy. First, in Chapter 9, he 'directs' a dramatic representation of pickpocketing, an inverted morality play performed by the Artful Dodger and Charley Bates, which Oliver perceives as 'a very curious and uncommon game' (p. 54). In Chapter 18, he watches behind the scenes as the Dodger, beer and tobacco in hand – believing himself to be the incarnation of 'romance and enthusiasm' (p. 116) – uses the capitalist vocabulary of self-help to persuade Oliver of the greatness of the life of the thief: 'Why, where's your spirit? Don't you take any pride out of yourself? Would you go and be dependent on your friends?' Then the young criminals enact two 'pantomimic representation[s]', one of 'a handful of shillings and halfpence', signifying a 'jolly life', and the other, Master Bates's party piece, of the gibbet. Money is thus made to seem attractive and immediate, while death by hanging is presented comically, thereby anaesthetizing its force as a real threat to life. The Dodger finally attempts to invert Oliver's moral values by telling him that he's been 'brought up bad' and that if he doesn't steal handkerchiefs and watches, 'some other cove will' . At this climactic moment, Fagin as director enters, advising Oliver to 'take the Dodger's word for it. [...] He understands the catechism of his trade' (Chapter 18, p. 118).

When these attempts fail, Fagin tells Oliver comic tales, oral narratives about crime that make Oliver laugh 'heartily [...] in spite of his better feelings'. They are the sugar to make the poison go down. The text makes explicit Fagin's consciousness of what we can call, literally, the arts of corruption:

> In short, the wily old Jew had the boy in his toils; and, having prepared his mind, by solitude and gloom, to prefer any society to the companionship of his sad thoughts in such a dreary place, was now slowly instilling into his soul the poison which he hoped would blacken it, and change its hue for ever.
> Chapter 18, p. 120

Fagin's intended master-stroke is to leave Oliver a volume which bears no accidental similarity to *The Newgate Calendar*. Fortunately and improbably, the book does not have the desired effect:

> It was a history of the lives and trials of great criminals; and the pages were soiled and thumbed with use. Here he read of dreadful crimes that made the blood run cold [...]. The terrible descriptions were so real and vivid, that the sallow pages seemed to turn red with gore [...].
> In a paroxysm of fear, the boy closed the book, and thrust it from him. Then, falling upon his knees, he prayed Heaven to spare him from such deeds.
> Chapter 20, pp. 129–30

The success that Fagin has had in persuading the Dodger and Bates of the 'greatness' of crime is everywhere evident, even in the Dodger's doodling; the Artful amuses himself 'by sketching a ground-plan of Newgate on the table with the piece of chalk' (Chapter 25, p. 158). But perhaps the most striking illustration of Fagin's manipulation of the myth of romantic criminality comes when the Dodger is captured by the police. Bates is so distraught that his friend has been caught for the mundane theft of a snuff-box that he comes very close to realizing that fame is rarely the lot of juvenile delinquents. What upsets him most is that the Dodger won't feature in the *Newgate Calendar*:

> 'Cause it isn't on the rec-ord, is it? [...] 'cause it won't come out in the 'dictment; [...]. How will he stand in the Newgate Calendar? P'raps not be there at all. Oh, my eye, wot a blow it is!'
> Chapter 43, p. 295

Fagin persuades Charley, however, that newspaper reports of the criminal trials will make the Dodger's name, acting out the court-room scene so brilliantly that Charley eventually sees his friend's capture as 'a game! a regular game!' (Chapter 43, p. 296).

When J. Hillis Miller wrote his influential essay 'The Fiction of Realism' (1971),[25] it was important to draw attention to the self-reflexive elements of *Oliver Twist* which had hitherto been neglected by approaches to the novel which privileged 'content' over 'form' and accredited little sophistication – aesthetic or ideological – to the young Dickens. However, Miller's vision of Dickens's universe, like much deconstructive work of its era, is nihilistic and perhaps slightly narcissistic; for Miller, Dickens's world 'creates illusion out of illusion and the appearance of reality out of illusion, in a play of language without beginning, end, or extra linguistic foundation'.[26] Miller uses his findings to support his epigraph, taken from *Sketches by Boz*, 'the illusion was reality itself'.[27] His wilful blindspot is that throughout the essay he takes the word 'illusion' to mean emptiness or hollowness, whereas in Dickens's fiction, theatrical and fictional 'illusions' are often intensely meaningful and 'real' to those engaged with them. And perhaps more importantly, Dickens's novels present not an anarchic free play of illusions, but a careful positioning of key discourses in which fictions, both textual and 'real', inevitably exert ideological (and moral) influence and are always part of a material chain of cause and effect. In this sense, they have much in common with Bentham's theory of fictions, which emphasizes the double nature of verbal fictions, which are both illusory and real.[28] Dickens exploits both possible meanings of the epigraph of *Sketches by Boz*, 'the illusion was reality itself'.

Since Miller's pioneering work, a number of critics have tried to take his analysis of the self-reflexive elements of *Oliver Twist* further, maintaining as I do that, in Steven Connor's words, 'there are ways of taking self-reflexivity seriously without drowning every instance of it in an undifferentiated ocean of textuality'.[29] I am talking here about D. A. Miller, Robert Tracy, Steven Connor himself, and Stephen Bernstein, all of whose analyses of *Oliver Twist* mention Newgate narratives particularly and explore Dickens's textual play with narratives and fictions more generally.[30] However, none of them places this discussion within the framework of the social and cultural shifts of the 1830s, in particular the beginnings of mass culture. Historicizing the text from this perspective brings an important detail into focus. Dickens is not just interrogating the nature of narratives and fictions in *Oliver Twist* (as some of this analysis of self-reflexivity, inspired by post-structuralism, is inclined to suggest); he is as concerned with the power of a variety of specifically *popular* cultural forms. When the novel self-reflexively alludes to the Newgate debate, Newgate myths are nearly always disseminated via cultural vehicles that are accessible to the lowbrow and highbrow alike. In the first example discussed above, Fagin directs an inverted morality play in which the Artful Dodger and Charley Bates attempt to teach Oliver how to pick pockets. In the second, the two boys act out a 'pantomimic representation' of the joys of money on the one hand and the horrors of the gibbet on the other; next, Fagin resorts to oral narrative, telling comic tales; then the Dodger sketches Newgate on a table-cloth; and when Master Bates is on the point of realizing that the romance of crime is a myth, Fagin draws on a variety of popular cultural modes, acting, narrating, conjuring up the Dodger's performances in the law court (a favourite source of entertainment for the Victorians) and the newspaper reports of the same to convince Master Bates of the Dodger's inevitable honour and fame.

It is conspicuous that the least visibly effective of all these attempts to corrupt using Newgate myths is the famous instance when Oliver is given the volume resembling the *Newgate Calendar*. When faced with the written word in the physically substantial form of a book, Oliver takes the book almost as a warning of its own power and falls down on his knees to pray. When indoctrination is disguised as pleasure, or entertainment, by contrast, Oliver is far more susceptible: he watches the pick-pocketing demonstration attentively as a 'curious and uncommon game'. The same language of play is repeated when Charley Bates, after various impersonations and narratives of the Dodger's heroism from Fagin, also sees his friend's capture and imminent court appearance as a game. Fagin's comic tales of crime come closest to corrupting Oliver, making him laugh 'heartily [...] in spite of all his better feelings'. Even the Dodger, a cooler customer than Oliver and Bates, '*amuses* himself' (italics mine) by drawing a plan of Newgate on the table. It is true that what distresses Bates initially when the Dodger is captured, is that there won't be a formal, written 'record' of his achievements in the *Newgate Calendar*; it is also true that Fagin's conviction that the newspapers will report on the trial heartens Charley. But what is significant is that it is Fagin's humour and his abilities as an actor and story-teller which move Charley from tears to laughter; what we could call the 'illegitimation' of the written word finally proves incidental.

If the text is read closely, what seems to be at stake here is the function of emotion and pleasure as tools of power: amusement, laughter and play are the ultimate ideological vehicles. In the context of the Newgate debate, this has several repercussions. First, the genres central to the Newgate controversy were romance and melodrama, both of which (and often the two combined) relied on the immediate stimulation of emotion and pleasure for their popular appeal. Second, the potentiality and dangers of the mass market for culture which was emerging in the 1830s is under scrutiny. Dickens seems to be exploring theories that had been put forward by W. J. Fox in the radical *Monthly Repository* in the 1830s, about the possibilities for emotion, pleasure and also drama as democratic vehicles of communication which would prevent the divorce of intellectual from mass culture.[31] Fox was the editor of the radical, Benthamite paper, *The True Sun*, for which Dickens worked in his early days as a journalist.[32] The question of the function of emotion and pleasure in the dynamics of mass culture also has acute relevance, of course, to Dickens's own art, whose enduring popularity and cultural centrality have remained in many ways inscrutable to academic enquiry. The main problem critics seem to have had from the 1830s to the present is in reconciling Dickens's accessibility with his 'greatness' as a novelist, F. R. Leavis most famously trying to reconcile his own contradictions by labelling Dickens a 'great entertainer'.[33] The reason Dickens causes so many problems for academics, I would argue, is that however much we may bemoan the splintering of 'popular' from 'high' culture, our intellectual landscapes have internalized this division; Dickens's texts, by contrast, have not. A related problem for literary critics is the fact that, despite the advent of psychoanalysis, our profession still privileges rational analysis and the written word and lacks a satisfactory range of discourses for exploring the literary and cultural importance of emotion and pleasure. Cultural studies has arguably been quicker than 'literary studies' to explore the implications of post-Freudian analyses of emotion and pleasure,[34] mainly because popular culture is so obviously dependent on the engendering of emotion and pleasure for its defining popularity (hence the disdain with which literary studies so often regards it).

Dickens obviously felt strongly that art should be inclusive, and that art which was pleasurable, entertaining and emotionally engaging would achieve this – essays such as 'The Amusements of the People'[35] and a novel such as *Hard Times* (1854) prove this beyond doubt. However, his dislike of the exclusivity of the literary élite and his desire to be popular do not necessarily suggest a belief in a fully democratic art of the kind W. J. Fox seeks to define. The implicit relevance of Fagin's role as 'a generalized Newgate novelist'[36] to Dickens's own art is no accident. To Dickens, the story-teller is in a position of power; the comic entertainer who can manipulate the reader's emotions and arouse pleasure is all-powerful. This power is not necessarily a bad thing *per se* as long as (a perennial problem) power does not fall into the wrong hands; in the early Victorian period, when truths and realities seemed to be up for grabs, to sell your own reality convincingly was arguably a radical and proactive act from middle-class writers who had never before had such power. If you don't put your ideological spin on the world then, to quote the Artful Dodger, 'some other cove will'. From the perspective of author/reader dynamics then, for Dickens, inclusive art is not necessarily democratic.

That Dickens analyses power and the power of narratives and fictions self-reflexively is perhaps a predictable insight given the current obsession of postmodernism and post-structuralism with self-reflexivity, narratives and fictions. In this intellectual climate, then, it is particularly important to remind ourselves first, exactly what kind of narrative Dickens was writing in *Oliver Twist*, and second, that narrative is not Dickens's only concern. Unusually, Dickens's narrator, in one of the few critical interpolations in the whole of the Dickens canon, goes to great lengths in *Oliver Twist* to define and defend his novelistic practice. In the famous 'streaky bacon' passage in Chapter 17, Dickens cites 'good, murderous melodramas' as the model for his own 'craft'. (He specifically does not use the term 'narrative'.)

IT is the custom on the stage: in all good, murderous melodramas: to present the tragic and the comic scenes, in as regular alternation, as the layers of red and white in a side of streaky, well-cured bacon. The hero sinks upon his straw bed, weighed down by fetters and misfortunes; and, in the next scene, his faithful but unconscious squire regales the audience with a comic song. We behold, with throbbing bosoms, the heroine in the grasp of a proud and ruthless baron: her virtue and her life alike in danger, drawing forth her dagger to preserve the one at the cost of the other; and, just as our expectations are wrought up to the highest pitch, a whistle is heard: and we are straightaway transported to the great hall of the castle. [...]

Such changes appear absurd; but they are not so unnatural as they would seem at first sight. The transitions in real life from well spread boards to death-beds, and from mourning weeds to holiday garments, are not a whit less startling; only, there, we are busy actors, instead of passive lookers-on, which makes a vast difference. *The actors in the mimic life of the theatre, are blind to violent transitions and abrupt impulses of passion or feeling, which, presented before the eyes of mere spectators, are at once condemned as outrageous and preposterous.*

As sudden shiftings of the scene, and rapid change of time and place, are not only sanctioned in books by long usage, but are by many considered as the great art of authorship: an author's skill in his craft being, by such critics, chiefly estimated with relation to the dilemmas in which he leaves his characters at the end of every chapter: this brief introduction to the present one may perhaps be deemed unnecessary. If so, let it be considered a delicate intimation on the part of the historian that he is going back to the town in which Oliver Twist is born.

Oliver Twist, Chapter 17, pp. 105–6 (italics mine)

Dickens's conception of novelistic progression is evidently quite different from that of a diachronic, linear chain of events, or for that matter, the evolutionary *Bildungsroman* that is often assumed to be Oliver's story; neither does this passage suggest a sophisticated play with narratives. Oliver is an allegorical pawn in a paradoxical double novel whose anti-narrative principle is as strong as its narrative impulse. As the 'streaky bacon' passage makes clear, 'alternation', 'changes', 'violent transitions' and 'sudden shiftings', all potentially disruptive of narrative flow, co-exist with that very desire to forge narratives (or, to echo Dickens's own terminology, histories). Crucial to Dickens's defence of his method here is an emphasis on 'abrupt impulses of passion and feeling', natural to those experiencing the emotion, anti-naturalistic to those observing. But what I am chiefly interested in here is the attempt to educate the reader into a partly Brechtian understanding of the function of emotion in life, drama, and by implication, the novel.

The central place of popular stage melodrama and its raw material, passion, in Dickens's conception and practice of the novel, is aesthetically and ideologically fascinating in its implications. Like Barthes's 'doubly perverse' subject, Dickens manages to achieve the (for Barthes) barely conceivable, to combine the 'text of pleasure' with 'the text of bliss'.[37] To Dickens, emotion as an artistic and political tool can be used to disrupt narratives that reassure, to defamiliarize, and to fragment stories. Alternatively or simultaneously, it can do just the opposite, satisfying primitive, monopathic desires for wholeness.[38] All depends on our angle of vision, whether we are immersed in, or watching, the passions. Interestingly and characteristically, Dickens's example of those immersed in passionate experience is taken from the theatre: they are 'actors in the mimic life of the theatre', a slippage which complicates any temptation we may have to paraphrase this passage thus: 'narrative distances; drama immerses'. The generic fluidity in Dickens's conception of the novel explains the comparative ease with which he can theorize and utilize the novel as a site of pleasure *and* bliss. For Barthes, bliss is non-verbal, but somehow textual and thus by definition beyond definition;[39] for Dickens, unfettered by Barthes's privileging of the 'textual' over the theatrical, the source of 'bliss' is less abstract – it is the theatre, a genre he subjects to a process of Bakhtinian 'novelization'.[40]

The *Twist* passage has more profound resonances perhaps in relation to Raymond Williams's concept of 'structures of feeling'.[41] Williams's attempt to extend analysis of the nature and workings of ideology from focusing on 'impersonal' systems of thought to include 'personal' experience – and indeed to examine the relationship between the two – is notoriously difficult to pin down. However, what is relevant to the *Twist* passage is the distinction Williams makes between personal emotion – which to Williams is thought experienced in the present – and 'ideological systems of fixed generality', which our thought processes experience as past, static and fixed. So-called 'subjective' or 'personal' feeling is so powerful, Williams argues, that ideological systems are 'relatively powerless' against it.[42] 'The basic error' of Marxism, Williams argues, is 'the reduction of the social to fixed forms'. The *Twist* passage seems to understand that in order for an author to have an impact on the reader, ideology must be experienced as all encompassing emotion; if it is not experienced as such, it appears 'outrageous and preposterous' and, as Williams suggests, ineffective.

The logical implication of this is, however, that detachment on the part of the reader may enable him/her to maintain the ability to analyse the text's ideologies. Such a detached reader is not, perhaps, the ideal reader Dickens had in mind, given his emphasis on the centrality of passion in aesthetic experience. It is initially tempting to speculate that if Dickens envisaged the perfect author–reader relationship as a marriage of equals, then perhaps the ideal reader, like the ideal author, would maintain the ability to immerse him/herself in emotional experience and to detach him/herself from the experience simultaneously. Interestingly, Dickens describes such an experience when he writes to Sir James Emerson Tennent on the experience of acting in *The Frozen Deep*. Dickens's description echoes the doubleness of the *Twist* passage, merging drama with the written word, public with private, and immersion with detachment, to evoke the ultimate aesthetic experience:

> As to the Play itself; when it is made as good as my care can make it, I derive a strange feeling out of it, *like writing a book in company. A satisfaction of a most singular kind, which has no exact parallel in my life. A something that I suppose to belong to the life of a Labourer in Art, alone, and which has to me a conviction of its being actual Truth without its pain,* that I never could adequately state if I were to try ever so hard.[43] (Italics mine)

Dickens's account of acting as 'like writing a book in company [...] actual Truth without its pain' is uncannily similar to Barthes's conceptualization of an aesthetic of textual pleasure (which encompasses both pleasure and bliss): Barthes concludes *The Pleasure of the Text*, 'If it were possible to imagine an aesthetic of textual pleasure, it would have to include: writing aloud. [...] A certain art of singing can give an idea of this vocal writing; but since melody is dead, we may find it more easily today at the cinema'.[44] For an author like Dickens who wrote novels 'in character'[45] and performed his own work publicly, such an aesthetic of 'textual' pleasure (as Barthes persists in calling it) was not theoretical but real. Tellingly, however, while Barthes's aesthetic is theorized from the viewpoint of an audience member, Dickens's is centred on the 'Labourer in Art, alone'. The audience/reader participates in Dickens's aesthetic of pleasure in only a secondary role. Ultimate pleasure and the power which attends it are the artist's – though the artist, like a benevolent patron, includes the audience (on which he depends) in the experience.

To argue that emotion is central to Dickens's popular art and indeed to the mechanics of popular culture is not then to say that his novels or the responses they are designed to elicit in the reader are straightforwardly or naïvely emotional. In the emotional economy of a Dickens novel, excess or release is inseparable from restraint – whether the restraint takes the form of the rigidly formulaic plot structures of melodrama, or the satirical or parodic narrative voice.[46] To talk generally, much Dickens criticism seems to me to be a variation on one of two themes – arguing either that Dickens is an intellectually sophisticated writer whose self-reflexivity prefigures postmodernism, or (less frequently in the present academic climate) that he is an emotional writer, a liberal humanist who uses emotion as a 'universal' language.[47] One emphasizes thought and systems, the other individuality and feeling. The reality is that both analyses are half true. Isobel Armstrong's definition of the Victorian 'double poem' as positing a 'content' as well as a self-reflexive critique of that content is useful here; it is 'an expressive model and an epistemological model simultaneously'.[48] For me, Dickens's novels are extreme examples of the 'double novel', a monopathic, melodramatic, child-like world of emotions and pleasures, co-existing with an inherently divided self-reflexive critique of the same. Similarly, the impulse to forge narratives, or connected chains of events, co-exists in Dickens with the anti-narrative, dramatic principle which celebrates the immediate pleasures of emotion and the body. In Dickens, doubleness does not take the form (as it does in Armstrong's conception of the double poem) of struggle or even dialogue; Dickens's is a doubleness so profound that it is unaware of itself as double.

Armstrong argues that the Victorian poem was able to achieve its doubleness only because Victorian poets were conscious of their 'secondary' position in Victorian mass culture.[49] Dickens disproves the potentially élitist assumption that profundity of insight into one's culture depends on being marginal to it. On the contrary, in *Oliver Twist*, the interrogation of the potential of popular culture as a vehicle of power, and indeed the infiltration of 'official' culture by popular cultural modes, is enabled not simply by abstract intellectual self-reflexivity, or endless narratives about narratives, but by his reliance on the variety of cultural modes he analyses: melodrama, court reports, romance, newspapers, Newgate narratives, etc. For Dickens, these are ideological and moral vehicles, as well as objects of intellectual analysis or cultural enquiry. And this, of course, is where the perceived difficulty enters academic debate. Because much cultural theory has taught us that many of these modes are conservative soporifics to the masses, the most urgent critical question is always: was Dickens condoning or criticizing the conservatism of these popular cultural forms? My view of Dickens's texts as profoundly double should make it obvious that my answer to this question is that Dickens both peddled and undermined this conservatism, often at the same time. His texts thus emphasize the liberating as well as the imprisoning potential of the popular cultural experiences of emotion and pleasure. But my main point is that the question of conservatism is not the only, or the most interesting, question we should be asking about ideology and genre, leading as it often does to for-and-against political tribalism. What seems to me a more difficult and neglected area in Victorian studies is the function of emotion and pleasure in the power dynamics of an emergent mass culture.[50]

Notes

Brief extracts from this essay have appeared in the introduction to Juliet John (ed.), *Cult Criminals: the Newgate Novels*, 6 vols (London: Routledge, 1998), I, v–lxxi. I am grateful to Avril Horner and Antony Rowland for their comments on an earlier draft of this essay.

1. William Makepeace Thackeray, 'Horæ Catnachianæ', *Fraser's Magazine*, 19 (April 1839), 407–24 (p. 408).
2. *The Newgate Novel, 1830–1847: Bulwer, Ainsworth, Dickens, and Thackeray* (Detroit: Wayne State University Press, 1963), pp. 14–15.
3. These novels have recently been reprinted by Routledge. See *Cult Criminals: The Newgate Novels*, ed. Juliet John, which includes these novels plus Bulwer's *Night and Morning* (1841), and a lengthy critical introduction.
4. John Sutherland, *The Longman Companion to Victorian Fiction* (Harlow: Longman, 1988), p. 462.
5. Hollingsworth, *The Newgate Novel*, pp. 145–8, p. 160; John Russell Stephens, *The Censorship of English Drama, 1824–1901* (Cambridge: Cambridge University Press, 1980), p. 66.
6. *Ibid.*, pp. 221–2.
7. *Oliver Twist*, ed. Kathleen Tillotson (Oxford: Clarendon, 1966). All references will be to this edition and will be shown in the text.
8. Hollingsworth, *The Newgate Novel*, p. 126; Kathryn Chittick, *Dickens and the 1830s* (Cambridge: Cambridge University Press, 1990), p. 127. See also Chittick, *The Critical Reception of Charles Dickens, 1833–1841* (New York: Garland, 1989).

9. 'Horæ Catnachianæ', p. 407.
10. (Oxford: Oxford University Press, 1948; repr. 1987).
11. *All the Year Round*, 18 (27 July 1867), 118–20 (p. 119).
12. See Lyn Pykett's article, 'The Real versus the Ideal: Theories of Fiction in Periodicals, 1850–1870', *Victorian Periodicals Review*, 15 (1982), 63–74, for a well-researched account of the critical debate between realists and idealists at the mid-century; Pykett argues correctly that their views were much more complex and less oppositional than literary history has subsequently assumed.
13. See Michel Foucault, *Discipline and Punish: the Birth of the Prison* (Harmondsworth: Penguin, 1991) (first publ. as *Surveiller et punir: naissance de la prison*, by Editions Gallimard, 1975; this trans. first publ. by Allen Lane, 1975; publ. in Peregrine, 1979; repr. Penguin 1991).
14. Michel Foucault's work sees the 'modern' period as rooted in the Enlightenment but actually arising from a self-consciously critical questioning of Enlightenment values first performed by Kant in his essay 'Was ist Auflärung?' (1784). Rejecting a positivist view of history, Foucault often refers to the modern period in necessarily vague terms as encompassing the last two centuries and originating at the turn of the eighteenth and nineteenth centuries. He is more precise when dating 'modernity'; Kant's 'Was ist Auflärung?' is for Foucault 'a point of departure: the outline of what one might call the attitude of modernity'; 'modernity', he defines 'rather as an attitude than as a period of history', an attitude encapsulated by the nineteenth-century writer, Baudelaire. See Foucault, 'What is Enlightenment?', in *The Foucault Reader*, ed. Paul Rabinow (Harmondsworth: Penguin, 1984), pp. 32–50 (pp. 38–9).

    Like Foucault, Isobel Armstrong identifies the 'modern' period with the emergence of a particular attitude rather than with specific historical events (although of course the two are related). For Armstrong, Victorian poets were the first writers to think of themselves as 'modern' and 'to be "new", or "modern" or "post-Romantic" was to confront and self-consciously to conceptualise *as* new elements that are still perceived as constitutive forms of our own condition' – *Victorian Poetry: Poetry, Poetics and Politics* (London: Routledge, 1993), p. 3.
15. Elliott Engell and Margaret F. King, *The Victorian Novel Before Victoria: British Fiction during the Reign of William IV, 1830–37* (London: Macmillan, 1984), p. 5.
16. *Ibid.*, pp. 29–33.
17. Chittick, *Dickens and the 1830s*, p. 24.
18. *Lucretia* (1846), in *Cult Criminals*, ed. John, pp. 297–334 (p. 314).
19. Ian Watt, *The Rise of the Novel: Studies in Defoe, Richardson and Fielding* (Berkeley: University of California Press, 1957); Nancy Armstrong, *Desire and Domestic Fiction: A Political History of the Novel* (Oxford: Oxford University Press, 1987). Armstrong's work discusses the novel and the middle class in terms of the discourse of sexuality.
20. Chittick, *Dickens and the 1830s*, p. x.
21. *Ibid.*, pp. 64, 87.
22. Hollingsworth's *The Newgate Novel* contains details of dramatizations of all the Newgate novels; see the introduction to this book for a brief outline of the history of the early-nineteenth-century theatre.

23. See for example M. J. Landa, *The Jew in Drama* (London: King, 1926); Lauriat Lane Jr., 'Dickens's Archetypal Jew', *PMLA*, 73 (1958), 94–100.
24. Barthes's essay, 'The Reality Effect', is reprinted in *Realism*, ed. Lilian R. Furst, Modern Literatures in Perspective (London: Longman, 1992), pp. 135–41, from *French Literary Theory Today*, ed. Tzvetan Todorov, trans. R. Carter (New York: Cambridge University Press, 1982), pp. 11–17.
25. See 'J. Hillis Miller on the Fiction of Realism', in *Realism*, ed. Lilian R. Furst, pp. 287–318 (repr. from 'The Fiction of Realism: *Sketches by Boz*, *Oliver Twist*, and Cruikshank's Illustrations', in *Dickens Centennial Essays*, eds Ada Nisbet and Blake Nevius (Berkeley: University of California Press, 1971), pp. 85–126).
26. 'The Fiction of Realism', p. 315.
27. From 'The Drunkard's Death', *Sketches by Boz* (Oxford: Oxford University Press, 1957; repr. 1987), p. 493.
28. See *Bentham's Theory of Fictions*, ed. C. K. Ogden (London, 1932); see also Isobel Armstrong's fascinating discussion of Bentham's theory in relation to Browning in her book *Victorian Poetry*, p. 146, pp. 148–54.
29. '"They're All in One Story": Public and Private Narratives in *Oliver Twist*', *Dickensian*, 85 (1989), 3–16 (p. 3).
30. D. A. Miller, *The Novel and the Police* (Berkeley: University of California Press, 1988); Robert Tracy, '"The Old Story" and Inside Stories: Modish Fiction and Fictional Modes in *Oliver Twist*', *Dickens Studies Annual*, 17 (1988), 1–33; and Robert Bernstein, '*Oliver* Twisted: Narrative and Doubling in Dickens's Second Novel', *Victorian Newsletter*, 79 (1991), 27–34.
31. See Isobel Armstrong's, *Victorian Poetry* for discussion of Fox's ideas.
32. Peter Ackroyd, *Dickens* (London: Guild Publishing, 1990), pp. 136–7.
33. F. R. Leavis, *The Great Tradition* (London: Chatto and Windus, 1948; Harmondsworth: Penguin, 1962), p. 29.
34. See Roland Barthes, *The Pleasure of the Text* (1973), trans. Richard Miller (Farrar, Straus and Giroux, 1975; Oxford: Basil Blackwell, 1990); Mikhail Bakhtin, *Rabelais and his World*, trans. Hélène Iswolsky (London: MIT Press, 1968); Michel Foucault, *The Use of Pleasure* (1984), trans. Robert Hurley in *The History of Sexuality*, 3 vols (Pantheon Books, 1985; Harmondsworth: Penguin, 1992) II; and Fredric Jameson, 'Pleasure: A Political Issue', in *Formations of Pleasure* (London: Routledge & Kegan Paul, 1983), pp. 1–14; see also Antony Easthope's relative discussion of pleasure in high and popular culture in *Literary into Cultural Studies* (London: Routledge, 1991), pp. 95–7.
35. *Household Words*, 1 (30 March 1850), 13–15 and (13 April 1850), 57–60.
36. Tracy, '"The Old Story" and Inside Stories', p. 20.
37. Text of pleasure: the text that contents, fills, grants euphoria; the text that comes from culture and does not break with it, is linked to a *comfortable* practice of reading. Text of bliss: the text that imposes a state of loss, the text that discomforts (perhaps to the point of a certain boredom), unsettles the reader's historical, cultural, psychological assumptions, the consistency of his tastes, values, memories, brings to a crisis his relationship with language.

   Now the subject who keeps the two texts in his field and in his hands the reins of pleasure and bliss is an anachronistic subject, for he simultaneously and contradictorily participates in the profound hedonism of all culture (which permeates him quietly under cover of an art de vivre shared

by the old books) and in the destruction of that culture: he enjoys the consistency of his selfhood (that is his pleasure) and seeks its loss (that is his bliss). He is a subject split twice over, doubly perverse.
Roland Barthes, *The Pleasure of the Text*, p. 14.
38. Robert Heilman defines 'monopathy' as 'the singleness of feeling that gives one the sense of wholeness' – *Tragedy and Melodrama: Versions of Experience* (Seattle: University of Washington Press, 1968), p. 85; such emotional structures usually dominate in melodrama rather than tragedy, Heilman argues.
39. Barthes argues in *The Pleasure of the Text* that 'pleasure can be expressed in words: bliss cannot' and '*criticism always deals with the texts of pleasure, never the texts of bliss*' (p. 21).
40. Bakhtin's theory of 'novelization' claims that the elasticity of the novel form 'infects' other genres: 'In the process of becoming the dominant genre, the novel sparks the renovation of all other genres, it infects them with its spirit of process and inclusiveness' – *The Dialogic Imagination*, ed. Michael Holquist, trans. Caryl Emerson and Michael Holquist (Austin: University of Texas Press, 1981), p. 7. It can also be argued, of course, that the theatre 'infects' the novel in the *Twist* passage and in Dickens's fiction generally.

Note that Barthes concludes *The Pleasure of the Text* by citing the cinema as a site which combines pleasure and bliss, though the experience is still termed 'textual' rather than 'theatrical'.
41. See Raymond Williams, *Marxism and Literature* (Oxford: Oxford University Press, 1977; repr. 1986), pp. 128–35.
42. *Ibid.*, p. 129.
43. *The Letters of Charles Dickens*, eds Madeleine House, Graham Storey and Kathleen Tillotson, The Pilgrim Edition, 9 vols (Oxford: Clarendon Press, 1965–), VIII (1995), p. 256 (9 January 1857).
44. *The Pleasure of the Text*, pp. 66–7.
45. For his daughter Mamie's famous description of Dickens acting out his novels in front of a mirror, see J. B. van Amerongen, *The Actor in Dickens: a Study of the Histrionic and Dramatic Elements in the Novelist's Life and Works* (London: Palmer, 1926), p. 256.
46. See John Kucich, *Excess and Restraint in the Novels of Charles Dickens* (Athens: University of Georgia Press, 1981) and Tore Rem, 'Melodrama and Parody: a Reading that *Nicholas Nickleby* Requires?', *English Studies*, 3 (1996), 240–54 for analyses of this economy.
47. See Steven Connor, '"They're All in One Story"' and Philip Davis, 'Victorian Realist Prose and Sentimentality' in Alice Jenkins and Juliet John (eds), *Rereading Victorian Fiction* (Basingstoke: Macmillan, 1999) for two of the best examples of these different tendencies.
48. Armstrong, *Victorian Poetry*, pp. 13–14 (p. 13).
49. *Ibid.*, p. 3.
50. John Kucich's work has made an invaluable contribution to our understanding of the relationship between power, emotion and pleasure as it is figured in the Victorian novel, though unfortunately for my argument, he tends to neglect or disparage popular cultural modes – see, for example, his critique of melodrama in *Excess and Restraint*, pp. 43–57.

10
'More interesting than all the books, save one': Charles Kingsley's Construction of Natural History

Francis O'Gorman

The middle years of the nineteenth century saw a huge rise of interest in natural history in England. Supported by its own periodicals and specialist publications, its own clubs and societies, and its own technical equipment for specimen collecting, preservation and display, enthusiasm for natural history spread with rapidity across the country and across classes. A particularly accessible form of knowledge which insisted on the pleasurable information to be gained from the everyday, the garden pond, the sea-shore, the field, natural history was habitually constructed in broadly Christian terms as an activity which was reverent and which expanded the practitioner's knowledge of God: a typical description of the natural historian as 'lover of nature' was that given in 1854 in the new *Natural History Review* as one 'who has at his heart the advancement of the study, and therein the glory of his Creator'.[1] Natural history customarily offered no threat to orthodox creationism and, with Gilbert White as a memorable precedent, some of its most significant publishing practitioners were ordained ministers. If not making specific theological claims, furthermore, much natural history presented itself as generally morally improving and often animal life was depicted as emblematic of human truths, as in Elizabeth Surr's *Sea-Birds, and the Lessons of their Lives* (c. 1876). The broadly didactic and morally aware aspect of natural history commended itself as particularly appropriate for children and a portion of natural history writing (1840–70) was aimed explicitly at them.

Today, the corpus of natural history writing from the mid-century provides a valuable window into the formation and circulation of some Victorian ideas about nature, and ways in which knowledge of nature was constructed in theological, political, moral, aesthetic and gender terms. Despite this, however, many texts and practitioners of Victorian natural history remain peculiarly understudied. Lynn Merrill, in her important book *The Romance of Victorian Natural History*, remarks that 'natural history has been overlooked in the glittering array of Victorian sciences'.[2] Even in studies of practitioners like G. H. Lewes, who was distinguished in other fields, his natural history work is only briefly mentioned. Lewes published works of philosophy, science, literary criticism and some fiction. He also published *Sea-side Studies at Ilfracombe, Tenby, the Scilly Isles, and Jersey* (1858), a work of marine study which appeared at a time when this branch of natural history was, in the words of David Elliston Allen, 'a national craze'.[3] The craze was partly the result of the perfecting of the aquarium principle – for which, Allen observes, no single individual really deserves the credit – which allowed marine animals and plants to be kept alive for lengthy periods of time out of the sea. Many books were published in response to this rising tide of interest. A small sample from the 1850s includes Anne Pratt's *Chapters of the Common Things of the Sea-side* (1850), Philip Gosse's *A Naturalist's Rambles on the Devonshire Coast* (1852), *The Aquarium* (1854) and *A Handbook to the Marine Aquarium* (1855), Philip and Emily Gosse's *Sea-Side Pleasures* (1853), the anonymous *Sea Anemones; or, Tanks and their Inhabitants* (1856), the *Synopsis of British Seaweeds* (1857) taken from work by W. H. Harvey, Lewes's *Sea-side Studies*, and Charles Kingsley's *Glaucus; or, The Wonders of the Shore* (1854–5).

Charles Kingsley (1819–75), Anglican divine, author of *Westward Ho!* (1855) and *The Water-Babies* (1863), and prominent campaigner for increased standards in public sanitation, was an enthusiastic natural historian who, as Canon of Chester Cathedral, founded a botany class in 1870 which attracted unprecedented numbers of students. *Glaucus*, his most substantial work of marine natural history, was first published in the *North British Review* for November 1854 as an article reviewing a selection of new books on sea life, including Gosse's *Rambles* and *The Aquarium*. It was published in adapted form as a separate volume the next year. In the following essay I discuss this text in detail, looking both at *Glaucus*'s individual characteristics and its exemplification of more widespread features of mid-nineteenth-century natural history. I investigate the languages appropriated by the text in its presentation of natural history as fortifying moral faculties; I analyse *Glaucus*'s theology of nature, together with its various appropriations of discourses of empire and its conception of the natural world as pleasurable artefact and exotic commodity. Its reflections on the moral sensibility of the natural historian as collector of living specimens and its deployment of discourses of masculinity are also examined. My account, which looks more analytically than Lynn Merrill's useful but descriptive approach in *The Romance of Victorian Natural History*, reveals the plural ways of representing the natural world, constructing the practice of natural history, and shaping the subjectivity of the natural historian, which are aggregated within a single volume.

Glaucus (I refer unless stated otherwise to the first 1855 edition) begins by addressing a holiday-maker going down 'to pass [his] usual six weeks at some watering-place along the coast' (p. 217).[4] Bored, tired and idle, the holidaymaker can only look forward to lounging around unproductively at the resort while his sons, likewise frustrated at the lack of things to do, shoot 'at innocent gulls and willocks, who go off to die slowly' (p. 217). This miserable scene, says Kingsley, could be dramatically improved if only the holidaymaker – and his sons, presumably – would take a serious interest in the natural history of their coastal environment and 'try to discover a few of the Wonders of the Shore' (p. 218). The text then embarks on a substantial justification of natural history in moral, aesthetic, theological and gender terms before coming, in the second half of the book, to a sustained description and discussion of some of the 'Wonders of the Shore' themselves. *Glaucus* describes what can be found on the English coast, in rockpools and, more spectacularly, by dredging the seabed. In the final section, Kingsley advises on the formation of a successful and interesting aquarium stocked with some of the marine life already described. Natural history is presented throughout as a moral activity which pleasurably reveals engaging fact but also strengthens personal virtue.

This broad narrative, focusing on a change from idle *ennui* to profitable action, attainment and moral advance, reveals the text framed by the powerful and culturally prevalent Victorian discourse of self-improvement, emphasizing the productive use of leisure time as variously fortifying the body, mind and soul. Indeed, the whole activity of natural history in the Victorian period is closely associated with the language of self-improvement, rooted in the Carlylean conviction of labour as a source of redemption.[5] Instead of being 'wrapt up [...] in their little world of vanity and self-interest' (p. 218), the holidaymakers in the heavily moralized scheme of *Glaucus* will find salvation by opening their eyes to the natural world and its lessons, and by toiling in the biblically-phrased 'vineyard' (p. 310) of natural history. The redemption of the eye, so precious to John Ruskin, is a strong theme in much Victorian natural history.[6] Ruskin, himself the author of some works of quasi-natural history, declared in *Modern Painters*, ((1856), III), that 'The greatest thing a human soul ever does in this world is to *see* something, and tell what it *saw* in a plain way. [...] To see clearly is poetry, prophecy and religion, – all in one'.[7] Kingsley presents natural history as a purification of the eye which involves these Ruskinian elements of poetry, prophecy and religion, just as his own language endeavours to depict the exactitudes of marine life as his eyes had seen them. For him, the self-improving activity of profitable, pleasurable labour will result in a redeemed eye which 'sees significances, harmonies, laws, chains of cause and effect endlessly interlinked' (p. 224) and where '*The Art of Seeing*' is the 'highest faculty' (p. 238; emphasis original).

This construction of natural history as the art of seeing is one of Kingsley's strategies of legitimation in *Glaucus*. Fashioning natural history as a form of knowledge dependent on the full, natural functioning not of a rare human power, but of the eye, Kingsley is emphasizing its accessibility, democratizing its cultural availability. But he is also helping to naturalize his construction of science and thereby subtly legitimate and authorize the values he embeds within it. Presenting the practice of natural history as something familiar, healthy and natural as the proper functioning of the eye, as essentially a matter of clear, unbiased sight, *Glaucus* is facilitating the transmission of the values and ideological structures operating within it. A different but related naturalizing strategy is deployed in the lecture 'Science' given at the Royal Institution in 1867. Here, Kingsley, in a building dedicated to the professional, institutionalized pursuit of science, emphasizes not the élite nature of empirical investigation but its accessibility, insisting that it requires merely the proper application of common sense. '[S]cientific method', he says, '[...] needs no definition; for it is simply the exercise of common sense', just as mathematics is 'the development of [...] conceptions of form and number which every human being possesses'.[8] Making scientific methodology no more than the exercise of a 'natural' human power, Kingsley is stressing the 'naturalness' of his conception of science itself, just as he does in fashioning natural history as the proper functioning of the eye.

Glaucus recurrently asserts the improving nature of practising natural history: natural history both requires and strengthens moral faculties. The text expresses this partly by appropriating a popular discourse of morality in Victorian England: medieval chivalry. The appropriation allows for the articulation of the moral claims of natural history in already culturally prestigious terms. The natural historian must be 'like a knight of old', says Kingsley (p. 238): 'gentle and courteous' (p. 238); 'brave and enterprising, and withal patient and undaunted'; 'of a reverent turn of mind [...] believing that every pebble holds a treasure'; and, finally, having a 'solemn and scrupulous reverence for truth' (p. 239). Without 'truthfulness', Kingsley solemnly declares, clinching his comparison, 'science would be as impossible now as chivalry would have been of old' (p. 239). Taking up a culturally familiar form of exemplary behaviour (which had been given particular currency only a year before *Glaucus* in Charlotte Yonge's enormously popular novel *The Heir of Redclyffe* (1853)), Kingsley's text commends the moral virtues that can be learnt from the patient, reverent and scrupulous practice of natural history and invests that practice at the same time with what is intended to be an appealing heroic glamour.

Natural history may require and impart a scrupulous, knightly concern for empirical truth, but it also needs a knightly faith. *Glaucus* describes the practice of natural history as fortifying the very faculty of faith, as strengthening the mental capacity to believe in an authority and in received truth, and reveals how natural history can train the mind to believe by emphasizing how faith in scientific authority is necessary to it. Though we may not understand the laws of nature, *Glaucus* says, we must believe what we are told by those wiser and more knowledgeable than ourselves. For instance, though we may not understand the connections in nature and be puzzled why natural forms apparently alike are not in fact related, we must have faith in the truths of science:

> You go down to any shore after a gale of wind, and pick up a few delicate little sea-ferns. You have two in your hand, which probably look to you, even under a good pocket magnifier, identical or nearly so. But you are told to your surprise, that however like the dead horny polypidoms which you hold may be, the two species of animal which have formed them are at least as far apart in the scale of creation as a quadruped is from a fish [...]. You must believe it; for in science, as in higher matters, he who will walk surely, must 'walk by faith and not by sight'. (p. 235)

The link between science and 'higher matters' is key. Kingsley's text commends natural history as an activity which requires and expands belief in scientific authority but, by strengthening the very faculty of faith, it also creates the right frame of mind for belief in spiritual truths. Practising natural history is, in Kingsley's construction, good preparation for the Christian life.

Implicit within the language of chivalry is an assumption that natural history is an acceptable masculine preoccupation, strengthening masculine virtues. Kingsley is anxious to assert this gendered claim against those who maintain that natural history is 'a mere amusement, and [...] a somewhat effeminate one' (p. 236), suitable only for 'effeminate or pedantic men' (p. 238). He promotes the knightly activity of looking at the wonders of the shore as assuredly male, constructing masculine virtues in culturally uncontroversial terms. *Glaucus* insists, for instance, on the need for physical strength in natural history ('our perfect naturalist should be strong in body' (p. 238)), repudiating any claim that the pleasurable study of nature is not for red-blooded, active and energetic men. The gendered character of *Glaucus*'s science is further emphasized in the fusion of natural history with the (masculine) language of Victorian sportsmanship as part of Kingsley's campaign to assert the manliness of the natural history (ad)venture. *Glaucus* presents a figure who is both huntsman and student:

> Happy, especially, is the sportsman who is also a naturalist: for as he roves in pursuit of his game, over hills or up the beds of streams where no one but a sportsman ever thinks of going, he will be certain to see things noteworthy, which the mere naturalist would never find, simply because he could never guess that they were there to be found. I do not speak merely of the rare birds which may be shot, the curious facts as to the habits of fish which may be observed, great as these pleasures are. (p. 225)

The text seeks to legitimate natural history as an appropriately masculine endeavour by associating it with an already culturally acceptable male activity, just as Kingsley's lecture 'The Study of Natural History' (1872) declares that the intellect needed for natural history is the same as that which makes 'a great military man'.[9] The moral implications of the sportsman-naturalist, it should be added, a figure who may seem perversely paradoxical to us, are present in other aspects of Kingsley's approach to the natural world, as will be seen, in particular, in his handling of the natural historian's moral responsibility as specimen collector.

Kingsley insists on manliness in *Glaucus* expressly to inhibit any notion that the pleasurable pursuit of natural history, which allows and privileges an affective response to the beauties of the natural world, might effeminize the subjectivity of the male. Kingsley is not alone in parrying the idea of natural history as merely an 'effeminate amusement', though others adopt different strategies to preserve its masculine credentials. John Lubbock begins *The Beauties of Nature* (1892) with a long introduction marshalling a host of eminent men – Aristotle, King Alfred, Humboldt, Keble, Ruskin, Tyndall, Darwin, Kingsley himself – who have persuasively written about the beauties of the natural world, assuring the reader, by weight of numbers, that an affective response to nature is acceptably male.[10] Ruskin's strategies in *Proserpina*, his book of flower studies, are more sophisticated. In the subtitle of the volume, *Proserpina* (1875–86) insists that its subject matter is inextricably associated with the male – *Studies of Wayside Flowers, while the air was yet pure among the Alps, and in the Scotland and England which my Father knew* – while in his 1874 'Introduction', Ruskin's self-construction affirms that botanical work has not compromised his masculinity because it has been pursued only in occasional moments snatched from a life of public and more strenuously masculine labour. Botanical work has been almost impossible since he was young, he says, because *'Blackwood's Magazine*, with its insults to Turner, dragged me into controversy' in 1842, and 'I have not had, properly speaking, a day's peace since'.[11] Even at the present, he adds, he has barely enough time to gather together his 'two or three' notebooks of 'broken materials' of flower study for he is constantly threatened by 'heavier thoughts and work coming fast on me'.[12] Ruskin projects himself as never having been able to give much attention to the business of botanical study because of the obligation to defend a man maligned and the burden of daily work. His botany has been possible only in the margins of a life of public duty, its incompleteness a sign that it has not distracted from the 'heavier' masculine work which even now threatens to 'end [it] altogether'.[13] Emphasizing that natural history has been pursued only in the spaces left in a busy public life, Ruskin finds a different way from Kingsley to contest the notion that nature study might be deleterious to masculinity.[14]

The manly practitioners of Kingsley's natural history not only learn facts but also grow morally. Similar claims were to be made for the practice of more professionalized science in the later years of the century, especially by those trying to improve the position of science in school and university curricula,[15] but the terms had already been set, as I argue in my conclusion, in the rhetoric of earlier natural history. If practising effective natural history fortified moral and manly strength, it also opened up, *Glaucus* maintained, the book of nature as

God's book. The theology of *Glaucus* presents nature as teaching human beings about God's creative power, as revealing God's bounty and invention, and as inspiring worship. The wonders of the natural world discovered by the natural historian always testify, *Glaucus* maintains, to the artistry of God the creator: everywhere 'the naturalist acknowledges the finger-mark of God, and wonders, and worships' (p. 225). Philip Gosse wrote in *The Aquarium* that 'The Holy Spirit has deigned to employ [natural science] in all ages as a vehicle of instruction to man':[16] *Glaucus*'s rhetoric prioritizes natural history as instructing human beings about natural law. True science, Kingsley writes, always leads to 'the living and permanent knowledge of living things, and of the laws of their existence' (p. 233). The natural historian will learn always of the permanent laws which govern the lives of all organic things and to which they are ceaselessly obedient. Kingsley's confidence in natural history revealing law 'so complex and so wonderful' (p. 218) is a pivotal feature of his sustained argument for the unity of Christianity with science.[17]

At points the text addresses other theological issues. It has no time for the theological implications of (pre-Darwinian) theories of species change. Even in the version published in *North British Review*, Kingsley assailed 'transmutation theories', declaring that they would mean God had changed his mind. There was no evidence for it, he declared, and 'species remain as permanent and strongly marked' throughout the natural world as they had been from the very moment of creation.[18] *Glaucus* also declares that nature, under entire and unquestioning obedience to God's law, and always giving 'most beautiful and diversified proofs of an adherence to a settled order' (p. 259), is therefore perfect (p. 292). This notion of the 'perfection' of nature, however, is referred to only casually in the text, and its theological implications are left entirely undiscussed. In the first essay of his *Prose Idylls* (1873), Kingsley was more speculative in his theology of nature, boldly querying the Cartesian divide between human beings and the natural world. He notes the assumption that birds act only by instinct rather than by reason, then reflects:

> [But the] imputation of acting by instinct cuts both ways. We, too, are creatures of instinct. We breathe and eat by instinct: but we talk and build houses by reason. And so may the birds. It is more philosophical, surely, to attribute actions in them to the same causes to which we attribute them (from experience) in ourselves. 'But if so,' some will say, 'birds must have souls.' We must define what our

own souls are, before we can define what kind of soul or no-soul a bird may or may not have. The truth is, that we want to set up some 'dignity of human nature;' some innate superiority to the animals, on which we may pride ourselves as our own possession, and not return thanks with fear and trembling for it, as the special gift of Almighty God. So we have given the poor animals over to the mechanical philosophy, and allowed them to be considered as only mere cunningly devised pieces of watch-work, if philosophy would only spare us, and our fine human souls, of which we are so proud, though they are doing all the wrong and folly they can from one week's end to the other.[19]

Kingsley questions the reasons for viewing other forms of life as soulless but the real thrust of the argument as it develops beyond this passage is to insist that human beings acknowledge their own souls, whether other species have them or not. As in *Glaucus*, *Prose Idylls* raises theologically suggestive ideas about the natural world but does not dwell at any length on their implications.

Glaucus asserts that the natural world may be read as revealing the finger of God but the text also constructs nature and human beings' relationship with it in other ways. A particularly revealing and culturally potent language deployed in these causes is that of Empire. Early in Kingsley's hugely successful novel *Westward Ho!*, published only a year after the *North British Review* version of *Glaucus*, the hero, Amyas Leigh, is portrayed, aged 15, standing on a cliff, looking out to sea. He is, at this point in his youth, Kingsley writes:

> a symbol, though he knows it not, of brave young England longing to wing its way out of its island prison to discover and to traffic, to colonize and to civilize, until no wind can sweep the earth which does not bear the echoes of an English voice.[20]

Such an overt imperial gaze, in Foucault's terminology, is anticipated in descriptions of human relations with the natural world in *Glaucus*. When Kingsley writes of the pleasures of natural history, he considers the delight of discovering new species. Unfortunately, he observes, there are very few species left in England *to* discover and thus many botanists, entomologists and ornithologists are dispiritedly 'bemoaning themselves like Alexander, that there are no more worlds left to conquer' (p. 230). Most natural historians must therefore accept, Kingsley continues, that they will discover nothing new in their endeavours though 'in giving up discovery, one gives up one of the highest enjoyments of natural history' (p. 230).

The chivalric discourse of virtuous and manly knighthood appropriated by the text certainly has implications of (violent) conquest embedded within it, but *Glaucus*'s deployment of an imperialist discourse foregrounds the conquest of nature explicitly. Kingsley's description of searching under boulders at the extreme low-water mark presents us, in all the dramatic immediacy of the present tense, with an example: 'Now, the crowbar is well under it; heave, and with a will; and so, after five minutes' tugging, propping, slipping, and splashing, the boulder gradually tips over, and we rush greedily upon the spoil' (pp. 263–4). The 'worlds left to conquer' for English natural history may not be, according to Kingsley, new species, but territory already discovered is, in this passage, literally 'despoiled' again. The natural historian is constructed as the conqueror of nature, as its subduer free to take his bounty as the reward for the labour of conquest.

Glaucus's imperialist discourse is augmented in the last section of the text where collecting for the aquarium is discussed, for the capturing, identifying, describing, and exhibiting required by the aquarium have colonial connotations. The natural historian's collection, the specimens of a natural world described in colonially revealing terms 'as full of inhabitants as [...] the Amazon or the Gambia' (p. 263), implicitly asserts the authority of the collector to appropriate and display 'foreign' ways of life, to signify superiority by disclosing his power to organize, describe and own examples of other forms of life. *Glaucus* pictures the 'live stock' (p. 300) of the aquarium in rich terms, presenting the specimens of the natural world as exotic treasure to be gazed at with wonder:

> If you find [a *Dianthus*, often found attached to live oysters], clear the shell on which it grows of everything else (you may leave the oyster inside if you will), and watch it expand under water into a furbelowed flower, furred with innumerable delicate tentacula; and in the centre, a mouth of the most brilliant orange; altogether one of the loveliest gems, in the opinion of him who writes, with which it has pleased God to bedeck His lower world. (p. 302)

This language fashions the specimen, now the 'property' of the colonialist-collector, as a coveted jewel from the 'perfect' but nonetheless 'lower' world, a gem found by the fortunate explorer and available for his taking. The natural world here provides the collector with valuable goods for his consumption in a culture where the commercial marketplace was increasingly demanding exotic products from around the world. *Glaucus* deploys the language of nature as exotic artefact recurrently, fashioning coastal natural history as an encounter with landscapes to wonder at and strange forms to marvel over.[21] There 'are along every sea-beach', Kingsley writes, 'more strange things to be seen, and those to be seen easily, than in any other field of observation which you will find in these islands' (p. 230). Kingsley dwells on the out-of-the-ordinary, the brightly coloured and fantastic; his eye lingers on a sea-slug 'of a bright lemon-yellow, clouded with purple, [...] another (exquisite little creature) of a pearly French white, furred all over the back' (p. 272); he recounts the 'delicious Italian climate' of Torquay with its 'endless variety of rich woodland, flowery lawn, fantastic rock-cavern' (pp. 246–7) and remembers the unpromising lump on the wet sand which, when poked, produced to his amazement 'a pair of astonished and inquiring horns and a little sharp muzzle' (pp. 254–5). The natural historian's business, according to *Glaucus*, will always bring him to such unexpected treasures, such exotic and surprising forms of life. Periodically, the strange forms are described in even more dramatic terms. 'Some, surely,' Kingsley writes:

> can recollect at their first sight of the Alpine Soldanella, the Rhododendron, or the black Orchis, growing upon the edge of the eternal snow, a thrill of emotion, not unmixed with awe; a sense that they were, as it were, brought face to face with the creatures of another world. (p. 231)

Such is nature's capacity to amaze that even the smallest microscopic animals, understood correctly, appear as impressive as 'those gigantic monsters, whose models fill the lake at the New Crystal Palace' (p. 233).[22] Any frisson of sensationalism produced by such exotic wonders, however, as encouraged most obviously in Victorian England by shows exhibiting freaks of nature, is studiously avoided. The only legitimate feeling Kingsley allows is 'a thrill of emotion, not unmixed with awe': the border dividing this 'thrill of emotion' from sensationalism is carefully policed by the religious connotations of 'awe'.

The discourse of the exotic, Renata Mautner Wasserman remarks, 'mediates between the defining self and a more radical otherness', and is itself an expression of power.[23] The exotic places the self in a position of authority over the 'foreign', controlling and shaping it; it emphasizes the distance of the other from the self while preserving its allure and readiness for consumption. Kingsley's construction of the natural world as exotic, as wondrous and captivating in its strangeness, is in part an expression of what Said calls *'positional* superiority';[24] it is a discourse which seeks to manage the otherness of, in this case, nature by dwelling on its curiosity and its bizarre and intriguing appeal in a way which also implies its ripeness for exploration/exploitation and its kinship with exotic commodities ready for consumption by a superior 'defining self'. The language of the exotic allows Kingsley a further way in which to situate human beings in authority over a consumable and exploitable natural world.

Glaucus's construction of nature as exotic is important in the text's leverage of power. But that power also brings with it a degree of responsibility. *Glaucus* maintains that the owner of the aquarium has a duty of care towards his 'pets' (p. 304) and the colonialist's discourse in this respect is not without the vocabulary of burden. If 'you leave your vase in a sunny window long enough to let the water get tepid', Kingsley writes, 'all is over' (p. 304). But the responsibility for keeping the specimens 'as happy and as gorgeous as ever' (p. 301) extends only some way. At times, *Glaucus* admits that the extraction of specimens from their natural environment is at the cost of their lives but that this is no argument to restrict it. Thus, the paradox of the sportsman-naturalist reappears in *Glaucus*'s approach to the morality of specimen collecting. In the following passage, the imperative language arises from the text's belief in the right of the natural historian to acquire for his own aesthetic pleasure even though his action brings death to the source of it. Vividly describing a particular episode on the beach, Kingsley writes:

> But there is one, at last! – a grey disc pouting up through the sand. Touch it, and it is gone down, quick as light. We must dig it out, and carefully, for it is a delicate monster. At last, after ten minutes' careful work, we have brought up, from a foot depth or more – what? A thick, dirty, slimy worm, without head or tail, form or colour. A slug has more artistic beauty about him. Be it so. At home in the aquarium (where, alas! he will live but for a day or two, under the new irritation of light), he will make a very different figure. (p. 255)

Even though the captured creature here is 'one of the rarest of British sea-animals, *Actinia chrysanthellum*' (p. 255), *Glaucus* does not admit any regret about its destruction save the momentary 'alas!'. The words of command ('We must dig it out'), which ironically would serve as well for more conventionally marketable riches mined from the ground, assert the collector's right to acquire and take pleasure over any other consideration. The drama of such acquisition is most vividly narrated in the long passages on sea-dredging from Gosse's *Aquarium* which Kingsley quotes (pp. 283-8).

Moral hesitations about collecting specimens, dead or living, from nature are infrequent in Victorian natural history. The language of conquest is usually more prominent than hesitations about destruction or death. Philip Gosse's *Natural History: Fishes* (1851) is typical of the prevailing attitude to animal death. He pointed out that killing specimens was acceptable because 'They have no terrors of futurity beyond death, and probably have little fear of death itself, beyond the habitual apprehension which prompts the exercise of caution and sagacity'.[25] Others distanced themselves from charges that they brought *cruel* death but did not question their general right to kill in the first place. Montagu Browne in *Collecting Butterflies and Moths* (c.1878), a hobby peculiarly targeted with charges of cruelty,[26] wrote:

> A vast amount of ignorant ideas, carefully nursed, are used as weapons against the entomologist – the pet one of which is, that impalement of a living insect through the head constitutes the sole aim and end of the collector. The fact is curiously inverse of this, for not only are insects captured for purposes of study, but they are never impaled alive but by a very ignorant or careless person. The lepidoptera [...] are very easy to kill, the simplest plan being to press the thorax underneath the wing with the finger and thumb, which instantly causes death. This is now superseded by the cyanide bottle, of which more anon.[27]

The Reverend Charles Alexander Johns, one of Kingsley's own teachers who is praised in *Glaucus* as a natural historian of 'most accurate and varied knowledge' (p. 309), similarly warned the readers of his book *Birds' Nests*, published the same year as the first version of *Glaucus*, not to do anything cruel – 'make a firm resolution to do nothing that your conscience tells you is cruel, either by destroying a nest, robbing it of its contents, or ill-treating the young birds'[28] – but nonetheless maintained the right to take and blow eggs for the purposes of making a good collection.

Other moral considerations in Victorian natural history are usually reminders of the collector's responsibility towards the welfare of live collections. Salutary warnings are given about negligence or thoughtlessness, usually pointed by gruesome details of the consequences of neglect. Children are the most usual audience. In one article in 1876 in the first periodical specifically for girls in the Victorian period, *Aunt Judy's Magazine*, readers were told a moral tale of death and carnivorous consumption in an aquarium kept by some children. J. H. Ewing's 'A Week Spent in a Glass Pond' ends with one of the children, contemplating the destruction of so much life, declaring: 'What makes me sorry is, that I don't think we ought to have "collected" things unless we had really attended to them, and knew how to keep them alive'.[29] Certainly, the end of *Glaucus* instructs the reader in his responsibility towards his collection, too, detailing how practically to make a safe and flourishing aquarium in which those creatures which can survive transplantation will. But, with a doubleness related to the tutelary figure of the sportsman-naturalist and echoing the moral postures of men like Johns, Kingsley allows little anxiety over the fact that extending the empire of the natural historian can occur at the cost of life itself.

Kingsley's text is the site for the simultaneous presentation of a range of discourses which construct natural history, the subjectivity of the natural historian, and human beings' relationship with the natural world in particular aesthetic, moral, theological, political and gendered terms. It is a rich example of a specific and popular practice of writing which both reflected and shaped fundamental ideas about nature and its study in mid-nineteenth-century England. *Glaucus*, however, is at the cusp of significant change. Cynthia Eagle Russett remarks that 'In the natural sciences the great event [of the nineteenth century] was the emergence of biology out of a union of descriptive natural history and physiology'.[30] But the new science of biology, in the wake of *The Origin of Species*, five years after the first publication of *Glaucus*, was to be dominated by evolutionary paradigms largely resisted by natural historians. To believe, however, that natural history thereafter rapidly lost its cultural authority is only to accept without challenge the model of hegemony advanced by the empirical scientists. Much more work needs to be done, in fact, in identifying and analysing the extensive presence, purchase and renegotiated authority of natural history, and its ways of perceiving the natural world, in various strata of society in *later* Victorian Britain.[31]

Natural history was obliged to reconfigure its authority, as the century moved on, in the face of increasingly institutionalized empirical science, but that professional science was to learn from the rhetoric and self-presentation of natural history and, in the later years of the century, to deploy legitimizing procedures not unrelated to the strategies of men like Kingsley. Where Kingsley had been anxious to assert the desirable and morally improving aspects of the practice of natural history, professional empirical scientists in the later decades of the century came increasingly to fashion a language which likewise presented the essential features of their methodology as conducive to moral and intellectual development. Of course, where natural historians had urged the Christian dimension of their inquiry into the natural world, the values espoused by later professionalized scientists were secular: science typically was held to provide a training in objectivity, in lucid and rational thinking, in the morally prestigious faculty of impartial consideration.[32] The values were different, but the claim that inquiry into the secrets of nature was of moral advantage to the practitioner links the rhetorical strategies of Victorian natural history at its apogee in the mid-century to those of the professional empirical scientists, as they sought to articulate and secure their cultural authority, in its later decades.

Notes

A version of this essay first appeared in *Worldviews: Environment, Culture, Religion*, 2 (1998), 1–15. I am grateful to Clare Palmer, her two anonymous readers, and to the editors of the present volume for their suggestions. Also thanks to Clare Pettitt.

1. *Natural History Review*, 2 (1854–5), [1].
2. Lynn L. Merrill, *The Romance of Victorian Natural History* (Oxford: Oxford University Press, 1989), p. 21.
3. David Elliston Allen, *The Naturalist in Britain: a Social History* (Harmondsworth: Penguin, 1978), p. 136.
4. All references to *Glaucus* given in the main text are to Charles Kingsley, *The Water-Babies and Glaucus* (London: Dent, 1908).
5. The *Natural History Journal*, begun in 1877, appropriately took as a motto for its title page Carlyle's 'LET EVERY MAN *find* HIS WORK AND *do* IT'.
6. See Merrill, *The Romance of Victorian Natural History*, p. 219.
7. *The Works of John Ruskin*, eds E. T. Cook and Alexander Wedderburn, 39 vols (London: Allen, 1903–12), V, 333.
8. Charles Kingsley, *Scientific Lectures and Essays* (London: Macmillan, 1885), p. 240; ibid., p. 241.
9. Ibid., p. 185.
10. See Sir John Lubbock, *The Beauties of Nature and the Wonders of the World We Live In*, 3rd edn (London: Macmillan, 1893), pp. [3]–38.
11. *The Works of John Ruskin*, XXV, 205.
12. Ibid.
13. Ibid.

14. For more on Ruskin and the discourses of masculinity, see Chapter 5 of my forthcoming *Late Ruskin* (Aldershot: Scolar).
15. See Frank M. Turner, 'Public Science in Britain, 1880–1919', *Isis*, 71 (1980), 589–608. The role of Ruskin's science in the debate about science and morals at the end of the century is discussed in my 'Ruskin's Science of the 1870s: Science, Education, and the Nation' forthcoming in *Ruskin and Late Victorianism*, ed. Dinah Birch (Oxford: Clarendon).
16. P. H. Gosse, *The Aquarium: an Unveiling of the Wonders of the Deep Sea* (London: van Voorst, 1854), p. 205.
17. For a discussion of some of Kingsley's problems in arguing for the unity of science with Christianity, see John C. Hawley, 'Charles Kingsley and the Book of Nature', *Anglican and Episcopal History*, 60 (1991), 461–79.
18. *North British Review*, 22 (1854), 24; There are further discussions of Kingsley's views of Darwin in Hawley, 'Charles Kingsley and the Book of Nature', 471; A. J. Meadows, 'Kingsley's Attitude to Science', *Theology*, 78 (1975), 20; and Owen Chadwick, 'Charles Kingsley at Cambridge', *Historical Journal*, 18 (1975), 313.
19. Charles Kingsley, *Prose Idylls: New and Old* (London: Macmillan, 1873), pp. 21–2.
20. Charles Kingsley, *Westward Ho! or, The Voyages and Adventures of Sir Amyas Leigh, Knight*, 3 vols (Cambridge: Macmillan, 1855), I, 18.
21. For the background to the conception of nature as artefact, see James G. Paradis, 'The Natural Historian as Antiquary of the World: Hugh Miller and the Rise of Literary Natural History', in *Hugh Miller and the Controversies of Victorian Science*, ed. Michael Shortland (Oxford: Clarendon Press, 1996), pp. 122–50.
22. In the gardens of the Crystal Palace were 'islands of irregular shape, covered with luxuriant vegetation [...] studded with the animals and reptiles of the antediluvian world' [Anon.], *The Crystal Palace Sydenham: Its History, Dimensions, Contents, and General Arrangements* (London: Adams, [n.d.]), p. 18.
23. Renata Mautner Wasserman, *Exotic Nations: Literature and Cultural Identity in the United States and Brazil, 1830–1930* (Ithaca: Cornell University Press, 1994), p. 14.
24. Edward Said, *Orientalism* (London: Routledge & Kegan Paul, 1978), p. 7; emphasis original.
25. P. H. Gosse, *Natural History: Fishes* (London: SPCK, 1851), p. 18.
26. See Allen, *The Naturalist in Britain*, pp. 145–7.
27. Montagu Browne, *Collecting Butterflies and Moths: Being Directions for Capturing, Killing, and Preserving Lepidoptera and their Larvæ* (London: 'The Bazaar', [1878?]), pp. [3]–4.
28. [C.A. Johns], *Birds' Nests* (London: SPCK, 1854), pp. 19–20.
29. *Aunt Judy's Magazine* (1876), [no vol. number] 733.
30. Cynthia Eagle Russett, *Sexual Science: The Victorian Construction of Womanhood* (London: Harvard University Press, 1989), p. 4.
31. There is a useful preliminary step in uncovering the status and purchase of later Victorian natural history in Bernard Lightman, '"The Voices of Nature": Popularizing Victorian Science', in *Victorian Science in Context*, ed. Bernard Lightman (Chicago: Chicago University Press, 1997), pp. 187–211.
32. See Turner, 'Public Science in Britain'.

11
Writing the Self and Writing Science: Mary Somerville as Autobiographer
Alice Jenkins

The career of Mary Somerville is a case-study in the interrelations of the branches and institutions of Victorian culture. Somerville was the foremost British Victorian woman of science; she earned a formidable international reputation for her scholarship, learning and ability to express clearly the most up-to-date and complex scientific and mathematical ideas. From the publication of her first major scientific work, *The Mechanism of the Heavens* (1831), through her hugely successful *On the Connexion of the Physical Sciences* (1834) and *Physical Geography* (1848), to her last book, published in 1869 when she was nearly ninety, *On Molecular and Microscopic Science*, Somerville's works were well received critically and – except for the last – were sold in astonishingly large numbers throughout the century.[1] The trajectory of Somerville's career is valuable in a reassessment of Victorian print culture, however, not just because of her critical and commercial success. In this essay I examine the nature of Somerville's work as a producer of texts, and focus largely on her *Personal Recollections*, published in 1873, shortly after her death.[2] My purpose is to explore the competing understandings of textual production which operate in this autobiography; to examine Somerville's role as a mediator of others' texts and to outline some of the ways in which her account of her own life was mediated by others after her death. Autobiography can be viewed as the most original, because the most isolating, of genres: the autobiographer is alone with only his or her life experience for material and a more or less strict notion of truth as the measure of success. As I explain later in this essay, though, feminist theorists in the last couple of decades have presented a different concept of autobiography, stressing the relational rather than the isolationist aspects of the genre. Neither objective accuracy nor originality of form is the measure of autobiographical success according to these accounts. In this essay I explore these questions of mediation, relation and originality as they were manifested in Mary Somerville's situation as a writer who was not male, not English, and yet an iconic figure in Victorian British culture.

Somerville's autobiographical writing includes at least two drafts of *Personal Recollections*. The final version of the book was edited after her death and published under the name of her elder daughter, Martha. In one of the draft passages which Martha chose not to publish, Somerville expressed disappointment with her own professional work. She attributed her sense of her own lack of originality as a scientist to the fact of her femaleness: 'I have perseverance and intelligence but no genius, that spark from heaven is not granted to the sex, we are of the earth, earthy, whether higher powers may be allotted to us in another existence God knows, original genius in science at least is hopeless in this'.[3] With her description of women as 'of the earth, earthy', Somerville alludes to Paul's letter to the Corinthians: Paul argues that the resurrected body will be made of a different substance from the flesh of the living body, just as 'there are also celestial bodies, and bodies terrestrial: but the glory of the celestial is one, and the glory of the terrestrial is another'.[4] For Somerville, then, women's connection with the natural world is too strong to allow for originality of thought *about* that natural world. The problem of the relationship between a woman scientist and her subject matter is one addressed repeatedly in the Victorian reviews of Somerville's work, which although almost uniformly favourable, used her femaleness to make points about the questions of what constituted science and what a scientist should be in this era of rapidly changing knowledge and practice. Somerville's role in the exploration of these questions, which – in their connections with the faith and doubt debate and with the development of technological and industrial power – were close to the heart of Victorian Britain's understanding of itself, has been largely ignored but is of considerable importance. The very word 'scientist', for example, was used in print for the first time in a review of Somerville's *On the Connexion of the Physical Sciences* in 1834.[5] In her own presentation of her books, and in the many textual and social networks which responded to them, Somerville is a neglected figure in the study of nineteenth-century British culture, perhaps because accounts of that culture have tended to concur with her own assessment of 'originality' as the primary measure of success. Somerville's scientific writing was founded on synthesis and comparison rather than on 'originality' in the heroic style.

Mary Somerville's first major publication was commissioned as a translation: a mediation of a text by a male, internationally recognized scientist. Apart from a small number of articles recording original scientific research, Somerville's career centred on textual mediations of others' work. She was not, however, a popularizer of science, like Jane Marcet, whose *Conversations on Chemistry* appeared in 1805 and had reached its thirteenth edition by Victoria's accession. Somerville's books were aimed at an adult, educated audience and her mediation usually consisted not of simplification and straightforward translation, but of explication, comparison and criticism. Besides commercial success, Somerville very quickly won international praise from the scientific establishment, represented both by individuals and by learned societies. Her authority on the scientific subjects on which she wrote was hardly questioned, but because her task was that of textual mediation rather than of physical experimentation, her authority partly derived from that of the scientists about whose work she wrote.

Somerville's work in scientific textual mediation and translation began when she was invited to translate the French scientist Pierre-Simon Laplace's monumental work on astronomy, *Traité de mécanique céleste* (1799–1825) for publication by the Society for the Diffusion of Useful Knowledge. The invitation came from Lord Brougham, who with Francis Jeffrey had founded the *Edinburgh Review* in 1802, and had been lambasted by Byron as one of the 'Scotch Reviewers' in *English Bards and Scotch Reviewers* (1809). Beginning a pattern that continued through much of her career, Brougham made his approach not to Somerville herself, but to her husband: 'I fear you will think me very daring for the design I have formed against Mrs. Somerville, and still more for making you my advocate with her; through whom I have every hope of prevailing' (quoted in *Personal Recollections*, p. 161).[6] Somerville reports herself to have been surprised 'beyond expression' at the invitation, and adds a comment which may be read as self-deprecating or, since it was written in the light of her subsequent success and celebrity, ironic: 'I [...] naturally concluded that my self-acquired knowledge was so far inferior to that of the men who had been educated in our universities that it would be the height of presumption to attempt to write on such a subject, or indeed on any other' (*Personal Recollections*, pp. 162–3). The book, in fact, went far beyond the original project of a translation. Somerville took on the task of explaining to the English reader the complex mathematics which underlay Laplace's work, and giving a rewriting, rather than a straightforward translation, of the original text. Indeed the original material she added was so substantial that Brougham found Somerville's completed work unsuitable for publication for the SDUK and the book was instead published by John Murray in 1831. Somerville's subsequent scientific career continued to be based largely on synthesizing rather than originating knowledge; but as Margaret Alic emphasizes, Somerville's books, though popular, were not popularizations. 'Rather she was an expositor, describing and explaining the current state of science in terms understandable to the well-educated reader.'[7]

Although unlike many other female scientists of her time she eventually gained comparatively easy access to the resources of nineteenth-century science, as a woman, a Scot and an auto-didact, Somerville was never able to enter unproblematically into its predominantly male, metropolitan world. A symbolic illustration of this was the inability of Somerville's friends, after her death, to have her buried in Westminster Abbey. The Victorian feminist and social activist Frances Power Cobbe attributed the refusal of the Astronomer Royal to nominate Somerville for this honour to 'jealousy, either scientific or masculine'.[8] Direct opposition to Somerville's work was voiced at various times during her life by members of institutions as powerful, but diverse, as the Church of England, the House of Commons, and her own family.

Further, her role as wife and mother made the beginning of her career particularly difficult: Somerville shared with other women writers the frustration of meeting social and familial demands at the expense of professional life. In her autobiographical writings she noted the different social estimation of male and female work: 'a man can always command his time under the plea of business, a woman is not allowed any such excuse' (*Personal Recollections*, pp. 163–4). The Victorian novelist Margaret Oliphant described her own early writing as 'subordinate to everything, to be pushed aside for any little necessity. I had no table even to myself, much less a room to work in, but sat at the corner of the family table with my writing-book, with everything going on as if I had been making a shirt instead of writing a book.'[9] Similarly, Mary Somerville records working in a public room, but unlike Oliphant she strove to keep her work private. Like Jane Austen she kept the fact of her writing hidden from her friends at first: 'there was no fire-place in my little room, and I had to write in the drawing-room in winter. Frequently I hid my papers as soon as the bell announced a visitor, lest anyone should discover my secret' (*Personal Recollections*, p. 164).

When Somerville's first book, *On the Mechanism of the Heavens*, appeared in 1831, her husband received a letter from William Whewell, a philosopher of science and also an eminent classicist. The letter enclosed a sonnet to Somerville, praising her book; Martha Somerville included the letter and sonnet in Mary's *Autobiography*. The sonnet is particularly interesting in a discussion of Mary's role in defining the relation of gender and science in this period. It opens with a specific allusion to Mary's sex – 'Lady' – and its closing lines describe her work as both intellectually and morally pure:

> thus we find
> That dark to you seems bright, perplexed seems plain,
> Seen in the depths of a pellucid mind,
> Full of clear thought, pure from the ill and vain
> That cloud the inward light!
>
> *Personal Recollection*s, p. 172

These lines do not make obvious the link between Somerville's gender and the clarity of mind that Whewell sees as a result of her freedom from professional vanity or controversy. But in his review of her second book, *On the Connexion of the Physical Sciences*, this link is made explicit. Almost all the reviews of this book discussed the fact of Somerville's femaleness: this was encouraged by the fact that the book was addressed specifically to women, and dedicated to Queen Adelaide. Whewell, like other reviewers, noted that Somerville's book would be of great use to men as well as to women, and concluded his review with a comment on the suitability of women to abstract thought:

> Notwithstanding all the dreams of theorists, there is a sex in minds. One of the characteristics of the female intellect is a clearness of perception, as far as it goes: with them, action is the result of feeling; thought, of seeing; their practical emotions do not wait for instruction from speculation; their reasoning is undisturbed by the prospect of its practical consequences. If they theorize, they do so
>
>> In regions mild, of calm and serene air,
>> Above the smoke and stir of this dim spot
>> Which men call earth.[10]

The unattributed quotation from Milton's masque *Comus* (1637), with which Whewell clinches his argument about the female mind, echoes his own sonnet in praise of Somerville's freedom from factionalism or professional jealousy. The lines are from the opening of *Comus*: a woodland spirit is explaining the difference between the lives of the 'bright aèreal Spirits' and the human beings who 'Strive to keep up a frail and Feaverish being' on Earth.[11] Although *Comus* adopts a classical rather than a Christian setting, this section of the masque, like the passage from Paul's epistle from which Somerville chose an epithet for herself, stresses the difference between earthly life and that which 'Vertue gives / After this mortal change, to her true Servants'.[12] Both Somerville's and Whewell's allusions use the contrast between the life of the earthly person and that of the resurrected or 'changed' individual as a metaphor for the difference between male and female mentalities. But where Whewell imagines Somerville, and all female thinkers, as represented by the 'purified' part of the analogy, Somerville sees herself as typical of the earth-bound part. So strong was her sense of her connection with the earth, indeed, that at the end of her autobiography, when at 88 she contemplated her life in heaven after death, she imagined herself missing the earthly world: 'I shall regret the sky, the sea, with all the changes of their beautiful colouring; the earth, with its verdure and flowers' (*Personal Recollections*, p. 348).

It seems strange that the paragraph in which Somerville describes women as 'of the earth, earthy' did not appear in the *Personal Recollections*. As an editor, Martha was at pains to emphasize her mother's domesticity, her dedication to her family, and her refusal to allow the 'recollections' to become either too personal (and gossipy) or too professional (and self-praising). Martha opens the *Recollections* with the discouraging remark that Somerville's life really fails to merit a memoir: 'the life of a woman entirely devoted to her family and to scientific pursuits affords little scope for a biography. There are in it neither stirring events nor brilliant deeds to record' (*Personal Recollections*, p. 1). Her mother's rather negative assessment of her own achievements would seem to agree with Martha's insistence on the modesty and lack of drama of Somerville's life. Martha, as editor, attempted to prevent her mother's text from expressing too strongly an individualist notion of the self, the autonomous and largely isolated selfhood which some major theorists of autobiography have seen as essential to the genre.[13] The argument has sometimes been made that even if the trajectory of the autobiographer's life has not been one of autonomy and individualism, the autobiographical narrator must necessarily structure the narrative by distinguishing the writing self from the lived self. By thus isolating the lived self from even its most intimate companion, autobiography isolates its subject from all other companions, including the reader. As Shari Benstock puts it in her critique of George Gusdorf's writing on the genre,

> from a Gusdorfian perspective, autobiography is a re-erecting of these psychic walls, the building of a linguistic fortress between the autobiographical subject and his interested readers [....] [T]he writing subject is the one presumed to *know* (himself) and this process of knowing is a process of differentiating himself from others.[14]

The modern feminist critics I discuss below have tended to re-evaluate the role of individualism in autobiography, and particularly by studying the autobiographies of women have generated a more relational model of the genre, in which networks, relationships and connections between the subject and both the people in her life and her readers are recognized as an essential aspect of the autobiography. The next part of this essay attempts to read Somerville's *Personal Recollections* in this way, and the attempt provides contrasting results for a relational model of autobiography.

Somerville's story of her life can be classed with Harriet Martineau's *Autobiography* (1877) or Francis Power Cobbe's account of her own *Life* (1894) as giving an account of the rise to professionalism of a Victorian woman; indeed both Martineau and Cobbe were among Somerville's friends, and it has been suggested that Cobbe may have had an influence on the editing of the *Personal Recollections*.[15] Cobbe may have been one of the 'valued friends, who think that some account of so remarkable and beautiful a character cannot fail to interest the public', whom Martha claims overcame her 'very great hesitation' at publishing her mother's memoirs (*Personal Recollections*, p. 1). Unlike Harriet Martineau, who felt even as a young woman that 'it was one of the duties of [her] life to write [her] autobiography', Martha claims on her mother's behalf to have required an outsider's urging to publish.[16] Cobbe herself is said by the editor of *her* autobiography to have required similar urging (Cobbe, *Life*, p. v). The recording of these conventional initiating moments in Victorian women's autobiographies seeks to establish the humility and modesty of the autobiographers. Linda H. Peterson notes a similar avowal of modesty accompanying the publication of Victorian women's diaries and journals. 'When published, an editor's note invariably preceded the text, disclaiming that publication was the author's motive. By mid-nineteenth century, this disclaimer had become a convention of the form'.[17] By 1899, when Margaret Oliphant's *Autobiography* was published, the convention of the initiating moment was still alive but was itself becoming a subject for autobiographical reflection. Oliphant's initiating moment is the comparison of her own life with that of George Eliot, which as she says 'stirred me up to an involuntary confession'.[18] Oliphant makes the conventional gestures of self-detraction in the introduction to her memoirs, but the modesty is mixed with self-pity and anger at her difficulties in life; this makes the disconcertingly painful initiating moment one of paradoxical self-assertion.

Somerville's autobiography is neither so emotionally intimate as Oliphant's, so sophisticated as Martineau's, nor so forthright as Cobbe's, but does share with the latter two in particular a strong political commitment to liberal causes and especially to the educational advancement of women. Politically, Somerville's views were in many ways in advance of her time. During the 1790s Somerville adopted liberal views in response to 'unjust and exaggerated abuse of the Liberal party', and maintained her liberalism to the end of her life. Even in extreme old age she wrote passionately in favour of animal rights, the abolition of slavery and the slave trade, female suffrage and the admission of women to university degrees. 'From my earliest years my mind revolted against oppression and tyranny, and I resented the injustice of the world in denying all those privileges of education to my sex which were so lavishly bestowed on men' (*Personal Recollections*, pp. 45–6). Her autobiography demonstrates, sometimes unconsciously, gender patterns operating in Victorian intellectual life: Somerville's professional career, for instance, was dominated by male readers, writers and friends, but her autobiography, while promising to focus on this part of her life, often emphasizes the 'female' spheres of domesticity, friendships with women, and the maternal role. This emphasis is not entirely, nor even largely, Somerville's own intention: the editorial presence of Martha Somerville makes itself felt throughout the *Personal Recollections,* attempting to set a tone of dignified feminine modesty and apologetically providing reminiscences of domestic incidents which her mother's narrative had ignored. Ironically, part of the modesty aimed for by Mary and imposed by Martha, involves silence about the intimately female parts of Mary's experience. Her unhappy first marriage, for example, which lasted three years, is dealt with in just five pages, of which one is made up of a letter of a much later date on the subject of tidal research, and another consists of Martha's factual interpolations about the family background of her mother's first husband. Mary's only direct published comment about her life with Samuel Greig outlines her dissatisfaction in the marriage:

> I was alone the whole of the day, so I continued my mathematical and other pursuits, but under great disadvantages; for although my husband did not prevent me from studying, I met with no sympathy whatever from him, as he had a very low opinion of the capacity of my sex, and had neither knowledge of nor interest in science of any kind.
> *Personal Recollections*, p. 75

It is typical of Somerville's outward-looking memoirs that she simply states the difficulties she encountered in her marriage, but does not reflect on the emotional effects of these difficulties.

Although Somerville's role in the home and family are repeatedly drawn to our attention by Martha, too much detail about these realms would have undermined the emphasis on her mother's femininity and dislike of 'gossip, and [...] revelations of private life' (*Personal Recollections*, p. 1). Martha makes this dislike of 'revelations' a founding principle of the autobiography, but if her mother did indeed dislike gossip, it was evidently only that which centred on her own family or close friends, since her autobiography gives numerous racy or outré stories about others – such as that of the Miss Boswell who to escape family-enforced seclusion 'eloped with her drawing-master, to the inexpressible rage and mortification of her father' (*Personal Recollections*, p. 56). Martha's task as editor, then, was to root her mother firmly in the realm of the domestic, while avoiding revealing much about the nature or structure of that domesticity. Estelle Jelinek sees this anxiety as typical of women's autobiographical writing: 'Even before Victorianism took hold, the impulse to intimate revelation was silent. Women continued to treat personal matters, but at a distance. To protect their vulnerable private lives, they wrote objectively about themselves and others'.[19]

Sometimes Somerville's interest in her family and community networks is hostile rather than affirmative. She allows herself some powerful moments of bitterness against the family, domestic and social structures which during the early part of her scientific career sometimes actively sought to prevent her from studying, and which more often simply failed to assist and support her. 'Unfortunately not one of our acquaintances or relations knew anything of science or natural history; nor, had they done so, should I have had courage to ask any of them a question, for I should have been laughed at. I was often very sad and forlorn; not a hand held out to help me', Somerville notes of her very early attempts to teach herself astronomy (*Personal Recollections*, p. 47). Later she laments her lack of a mentor to focus her studies:

> I continued my diversified pursuits as usual; had they been more concentrated, it would have been better; but there was no choice; for I had not the means of pursuing any one as far as I could wish, nor had I any friend to whom I could apply for direction or information. I was often deeply depressed at spending so much time to so little purpose.
>
> *Personal Recollections*, p. 72

As a result of these moments in which she expresses a combination of isolation remembered from the past and indignation felt in the present, Somerville's narrative presents her rise to fulfilment and success as an *individual* victory against a hostile or uncaring society. At these times it fits the model which sees autobiography as rooted in 'a defense of individual integrity in the face of an otherwise multiple, confusing, swarming, and inimical universe'.[20]

But Somerville's *Personal Recollections* cannot fit such a model in all respects. Her second marriage introduced her to a sophisticated circle of scientific and literary men and their wives, based primarily in London but also on the Continent. Mary's opportunities for serious scientific study seem to have become almost at once far wider and easier than most Victorian women and many Victorian men could have hoped for, as a result of this partly social, partly professional contact. After this part of her autobiography, a large section of most chapters until near the end of the book are taken up with chatty and rather inconsequential accounts of the brilliant parties, dinners and correspondence the Somervilles enjoyed with both scientific and social élites in Italy, France and Switzerland, as well as in Britain. Despite mysterious hints about financial troubles which necessitated thriftiness, Mary gives few further signs of conflict between her aspirations and her opportunities. It is as if the struggle for autonomy and individuality ceased when she entered a congenial marriage and her aspirations could be fulfilled.

This move towards returning the presence of the familial to a narrative which has deliberately sought to conceal it might be applauded by revisionist critics of autobiography, such as Susan Stanford Friedman, who argues that 'individualistic paradigms of the self ignore the role of collective and relational identities in the individuation process of women and minorities'.[21] But Martha's emphasis on her mother's success in the traditional feminine roles conflicts with, and apologizes for, Mary's attempt to develop, both in her practice and in her writing, *new* roles for women. Towards the end of the book, for instance, Somerville commented that

> Age has not abated my zeal for the emancipation of my sex from the unreasonable prejudice too prevalent in Great Britain against a literary and scientific education for women. [...] I joined in a petition to the Senate of London University, praying that degrees might be granted to women; but it was rejected.
> *Personal Recollections*, pp. 345–6

Martha immediately follows this passage with one of her bracketed, small-font editorial comments, which begins by appearing to consolidate both her mother's views and her authority in giving them, but which then goes on to repeat the movement of reclamation to the domestic which constitutes her own chief editorial effect:

> My mother, in alluding to the great change in public opinion which she had lived to see, used to remark that a commonly well-informed woman of the present day would have been looked upon as a prodigy of learning in her youth, and that even till quite lately many considered that if women were to receive the solid education men enjoy, they would forfeit much of their feminine grace and become unfit to perform their domestic duties. My mother herself was one of the brightest examples of the fallacy of this old-world theory, for no one was more thoroughly and gracefully feminine than she was, both in manner and appearance; and, as I have already mentioned, no amount of scientific labour ever induced her to neglect her home duties.
> *Personal Recollections*, pp. 346–7

Martha makes no attempt to challenge the hierarchy implicit in the 'old-world theory', by which domestic work is ranked more highly in a woman's achievements than her intellectual or professional life.

This emphatic location in the domestic sphere can easily be read as a defusing of the potentially dangerous textual presence of an intellectual woman. Elizabeth Winston finds in women's autobiographies published before 1920 a 'rhetoric of justification', used to 'establish a conciliatory relationship to their readers, by this means attempting to justify their untraditional ways of living and writing so as to gain the audience's sympathy and acceptance'. Winston cites the example of Lady Morgan's attempt, in her 1862 *Memoirs*, to portray herself as '*every inch a woman*'.[22] But an emphasis on domesticity and fulfilment of perceived female social and biological roles was sometimes made for less straightforward motives. For instance, Frances Power Cobbe, herself a powerful Victorian feminist and controversialist, adopted this femininity motif in her writing on Somerville. At one stage in her autobiography, Cobbe states that 'Mary Somerville was the living refutation of all the idle, foolish things which have been said of intellectual women. There never existed a more womanly woman' (Cobbe, *Life*, p. 386). We might read this statement, coming from Cobbe, as a necessary corrective to contemporary prejudices. But elsewhere in this book, Cobbe's attitude to Somerville is more complicated than this. One of the chief complications is the generation gap between the two women. Mary Somerville was already 42 when Cobbe was born. When they became friends in the late 1850s, Cobbe was in her mid-thirties and Somerville her late seventies. Cobbe presents her friend as very much part of a bygone, and unsophisticated, era: 'our fathers were in many respects, like children compared to ourselves' (Cobbe, *Life*, pp. 172–3). The difference of age operates in Cobbe's text to produce an oscillation between the two parts of the parent–child relationship: Cobbe's description pictures Somerville as both parent and child, vulnerable yet attractive in this dual role: Somerville was 'the dearest old lady in all the world, who took me to her heart as if I had been a newly-found daughter, and for whom I soon felt such tender affection that sitting beside her on the sofa [...] I could hardly keep myself from caressing her' (Cobbe, *Life*, pp. 383–4). As the Good Mother and at the same time the Good Daughter, Somerville is made to fulfil two of the three domestic roles of the Victorian woman.

Margaret Oliphant, born six years after Cobbe, also identified Somerville as a maternal figure, and she too goes on to turn this mother into a daughter, though not, as with Cobbe, her own daughter, but rather herself *as* daughter. Oliphant reads the account of childhood in *Personal Recollections* as reminiscent of both her mother's experience and her own: she read Somerville's autobiography shortly after it was published, and noted that it 'recalls to me my mother and even my own recluse childhood in the most delightful way'.[23] This similarity of childhood experience across a generation gap is associated with the distinctively Scots upbringing which both women described, and which I address in more detail below.

Valerie Sanders, in her study of Victorian women's autobiography, argues that childhood represented a safe space within a narrative potentially lacking in feminine modesty for women writers to express their discontent with the restrictions placed on children. (These restrictions might be thought of as metaphors for the difficulties faced by the adult woman in Victorian society.) 'Fraught with dangers as it was, the writing of childhood reminiscences nevertheless provided Victorian women with an opportunity to protest against the values and priorities of their society'.[24] In Somerville's published account of her childhood, she deals comparatively little with her development in terms of gender, focusing instead on her national identity as a Scot. At first she outlines the local customs she grew up with in her home in Burntisland, north of Edinburgh. She prefaces her account of the weddings, funerals and birth celebrations of the region with the remark that they were 'exceedingly primitive', but then provides a detailed and vivid description of these customs, frequently using dialect terms, which she usually signals with italics or inverted commas, emphasizing the difference both in time and in social standing between her language and that of her childhood neighbours.

This question of Somerville's Scottishness, particularly focused on her Scottish accent and use of 'aberrant' English, resurfaces from time to time throughout the text, and Somerville's attitude to it is ambiguous. It is also woven into her gender identity. Early in her *Personal Recollections*, for instance, Somerville explains that her mother taught her to read using the Bible (p. 17), but that when her father returned from his naval career and examined the state of Mary's education, he discovered that she 'read very badly, and with a strong Scotch accent; so, besides a chapter of the Bible, he made me read a paper of the "Spectator" aloud every morning, after breakfast' (p. 20). Her father's efforts to modify her accent were evidently not wholly successful, since Martha comments much later in the book that 'in speaking she had a very decided but pleasant Scotch accent, and when aroused and excited, would often unconsciously use not only native idioms, but quaint old Scotch words' (p. 120). This interest in Somerville's accent is material to an exploration of her role in Victorian intellectual society, since it serves as a reminder of her links with the eighteenth-century tradition of scientific and rational learning of the Scottish Enlightenment and with the world of the *Edinburgh Review* rather than with the London circles in which she did most of her scientific work, and also because it intersects with her position as a woman in a predominantly male professional world. Her reported description of herself as 'only a self-taught, uneducated Scotchwoman' recognizes the mingling of her national and gender differences from her colleagues; and this mingling was also recognized by her contemporaries. The Norfolk-born, Cambridge-educated, London-based chemist William Wollaston, for example, commenting on a draft of one of her early texts, took issue with a particular expression and remarked 'I know what you mean, but the term is either Lady-like or Scotch'.[25]

Somerville's account of her Scottish childhood was praised by Margaret Oliphant, herself another Scot. Oliphant was 'enchanted' with Somerville's story of her life and in writing her own, invoked Somerville as a kind of role model for older women. The first page of Oliphant's *Autobiography* cites Somerville's feeling that 'so far from feeling old, she was not always quite certain (up in the seventies) whether she was quite grown up!' and adds: 'I entirely understand the feeling'.[26] In both women's autobiographies, the sense of Scottishness is strongly associated with memories of the mother. Oliphant describes her father as 'a very dim figure' and her mother as 'all in all'. Her mother was 'of the old type of Scotch mother, not demonstrative, not caressing, but I know now that I was a kind of idol to her from my birth' (Oliphant, *Autobiography*, pp. 11–12). Somerville's father was English, and often absent on naval duty; the primary influence on her upbringing was her mother, whom she describes as very religious but also superstitious and fearful. Throughout her autobiography, indeed, Somerville associates Scottishness with outmoded fears and opinions: her highest praise of her second husband was that he was 'extremely handsome, had gentlemanly manners, spoke good English, and was emancipated from Scotch prejudices' (*Personal Recollections*, p. 87). Both Somerville's and Oliphant's autobiographies refer to nationality as a major constituent of identity; they exhibit a strong sense of the difference between a Scottish upbringing and an English one of the same period. This emphasis on nationality perhaps provides evidence against any conception of autobiography which is overly determined by the gender of the autobiographer. One of the few autobiographies Somerville mentions in her own *Personal Recollections*, for example, is that of David Brewster, the Scottish physicist. She introduces Brewster as a parallel to herself, in that he was born, like her, in Jedburgh, was only a year younger than she, and had experienced many of the same influences in his upbringing as she herself had. Somerville remarks 'few people will now understand me if I say I was *eerie*, a Scotch expression for superstitious awe', but notes that Brewster was one of those few, since he himself admitted to being 'eerie when sitting up to a late hour in a lone house that was haunted' (*Personal Recollections*, pp. 65–6). Brewster's autobiography gives Somerville a sense of connection with her own past, distant both in time and space, just as Somerville's *Personal Recollections* was to give Margaret Oliphant a link with her own remote (in both senses) childhood. Any account of Somerville's autobiography must take into account these national connections, which as I have indicated are often interlinked with gender but which can equally be (as in this instance) independent of it, and no less powerful for that.

Somerville's first major scientific work appeared a few years before Victoria's accession and she was revising her books for new editions until shortly before she died in 1872. Her writing career, though not her life, is thus roughly contemporary with that of Dickens. She moved in literary as well as scientific circles, and her sympathies were much more with the writers of her youth than with those of the years of her fame: Sir Walter Scott and Maria Edgeworth were among her friends. Unlike many of the novels by these two, though, Somerville's writing has not lasted into this century. Partly this is a result of genre: as scientific texts become outmoded they fade into the realm of history of science; Scott's and Edgeworth's novels can still be read for their intrinsic interest, as well as for their place in literary and social history. In addition, though, the aesthetic standard of 'originality', which she disappointedly applied to herself, has worked against Somerville's lasting scientific reputation outside specialist circles. If her major work were instead read using a standard which valued synthesis and collaboration, her enormous achievements might be more fully recognizable. After all, her *Personal Recollections*, as I have argued in this essay, represent the mediation by others of Somerville's autobiographical writings, and it is partly in the nature of this mediation that their use in a study of Victorian culture lies.

Notes

I am grateful to Ursula Tidd, Department of French, University of Salford, for her comments on an earlier draft of this essay.

1. Elizabeth C. Patterson notes that sales of the first nine English-language editions of *On the Connexion of the Physical Sciences*, for instance, exceeded 15,000 copies, 'remarkable for a scientific book': 'Mary Somerville', *British*

Journal for the History of Science, 4 (1969), 311–39 (p. 322). Patterson's *Mary Somerville and the Cultivation of Science, 1815–1840* (The Hague: Nijhoff, 1983) is the fullest study of Somerville's early scientific career.
2. Martha Somerville, *Personal Recollections, From Early Life to Old Age, of Mary Somerville* (London: John Murray, 1873). All further references are made in the text by page number to this edition.
3. Quoted in Patterson, 'Mary Somerville', p. 318.
4. 1 Corinthians 15. 40; 'of the earth, earthy' is in verse 47.
5. See, for example, Margaret Alic, *Hypatia's Heritage: a History of Women in Science from Antiquity to the late Nineteenth Century* (London: Women's Press, 1986), pp. 186–7.
6. Patterson notes that 'Much of Mrs. Somerville's scientific correspondence, especially in the period 1817–1840, was carried on through Dr. Somerville'. See Patterson, 'Mary Somerville', p. 318, n. 40.
7. Alic, *Hypatia's Heritage*, p. 188.
8. *Life of Frances Power Cobbe, As Told by Herself*, intro. by Blanche Atkinson (London: Swan Sonnenschein, 1904), p. 385. All further references are made in the text by page number to this edition.
9. *Autobiography and Letters of Mrs Margaret Oliphant*, ed. Mrs Harry Coghill (Leicester: Leicester University Press, 1974), p. 23.
10. [William Whewell], review of Mary Somerville, *On the Connexion of the Physical Sciences*, Quarterly Review, 51 (1834), 54–68 (p. 65). See John Hedley Brooke, 'Does the history of science have a future?', *British Journal for the History of Science*, 32 (1999), 1–20, for a fuller discussion of Whewell's review.
11. John Milton, *A Mask Presented at Ludlow Castle, 1634*, in John Milton, *The Complete Poems*, ed. B. A. Wright, 2nd edn (London: Dent, 1956; repr. 1980), pp. 47–75 (p. 53, ll. 3, 8).
12. *Ibid.*, p. 53, ll. 9–10.
13. Georges Gusdorf's highly influential argument in 'Conditions and Limits of Autobiography', in *Autobiography: Essays Theoretical and Critical*, ed. James Olney (Princeton: Princeton University Press, 1980), pp. 28–48, sees this concept of the individual, singular self as crucial to autobiography.
14. Shari Benstock, 'Authorizing the Autobiographical', in *The Private Self: Theory and Practice of Women's Autobiographical Writings*, ed. Shari Benstock (Routledge: London, 1988), pp. 10–33 (pp. 15–16).
15. See Patterson, 'Mary Somerville', p. 337.
16. Harriet Martineau, *Autobiography*, 2nd edn, 3 vols (London: Smith, Elder, 1877), I, 1.
17. Linda H. Peterson, *Victorian Autobiography: the Tradition of Self-Interpretation* (Yale University Press: New Haven, 1986), pp. 124–5.
18. Oliphant, *Autobiography*, p. 5.
19. Estelle C. Jelinek, *The Tradition of Women's Autobiography: from Antiquity to the Present* (Twayne: Boston, 1986), p. 41.
20. James Olney, *Metaphors of Self: the Meaning of Autobiography* (Princeton: Princeton University Press, 1972), p.15; quoted in Susan Stanford Friedman, 'Women's Autobiographical Selves: Theory and Practice', in Benstock (ed.), *The Private Self*, pp. 34–62 (p. 36).
21. Friedman, 'Women's Autobiographical Selves', p. 35.

22. Elizabeth Winston, 'The Autobiographer and Her Readers: From Apology to Affirmation', in Estelle C. Jelinek (ed.), *Women's Autobiography: Essays in Criticism* (Bloomington: Indiana University Press, 1980), pp. 93–111 (pp. 93, 96).
23. Oliphant, letter of 12 March 1874 to Blackwood, in *Autobiography*, p. 244.
24. Valerie Sanders, *The Private Lives of Victorian Women: Autobiography in Nineteenth-Century England* (Hemel Hempstead: Harvester Wheatsheaf, 1989), p. 54.
25. Quoted in Patterson, 'Mary Somerville', p. 332.
26. Oliphant, *Autobiography*, p. 3. 'Enchanted' appears in the letter to Blackwood, *Autobiography*, p. 244.

12
A Taste for Change in *Our Mutual Friend*: Cultivation or Education?

Pam Morris

At the climax of the main plot in *Our Mutual Friend* (1865), Mr Boffin gives voice to an extraordinary explosion of assumed rage at John Rokesmith or Harmon:

> 'Win her affections,' retorted Mr Boffin, with ineffable contempt, 'and possess her heart! Mew says the cat, Quack-quack says the duck, Bow-wow-wow says the dog! Win her affections and possess her heart! Mew, Quack-quack, Bow-wow!'[1]

Later, Mr Boffin claims this as his finest moment: 'I couldn't tell you how it come into my head or where from, but it had so much the sound of a rasper that I own to you it astonished myself'(*OMF*, Book IV, Chapter 13, p. 756). Mr Boffin is being a shade disingenuous here. Clearly he has been attending lectures on 'The Science of Language' given at the Royal Institute from 1861 to 1863 by Max Müller. In these lectures Müller expounded the 'Bow-wow' theory of language which he also illustrated by way of quack-quacks and mew-mews.[2] This was Müller's way of ridiculing theories of language that attempted to trace the origins of words back to imitation of natural sounds. Müller's own theory of origins, influenced by 'the genius of Darwin', located the development of language in two simultaneous processes: phonetic decay and dialectical regeneration (*Lectures*, II, 305; I, 40). Müller locates the principle of linguistic vitality and change in non-standard forms of expression, in dialects of all kinds. He argues that centralization, cultivation and especially literary languages perform the necessary function of stabilizing meaning so that language can acquire 'that settled character which is essential for [...] communication' (Müller, *Lectures*, I, 59). However, at a certain point this process of cultivation leads to over-formality; language stagnates, becoming like the frozen surface of a river 'brilliant and smooth, but stiff and cold' (Müller, *Lectures*, I, 57). Below this surface, the vital currents of popular dialectical language still flow freely and when centralized decay has reached a certain point of crisis they will rise up from beneath 'and sweep away, like the waters in spring, the cumbrous formation of a bygone age' (Müller, *Lectures*, I, 57). Müller believed that such moments of linguistic regeneration are most likely to occur as political commotions when the higher classes are either crushed in religious or social struggles or mix again with the lower classes (*Lectures*, I, 57).

Müller's lectures attracted large and distinguished audiences; J. S. Mill and Tennyson were among those who attended. Müller was extensively reviewed and quoted in the *Edinburgh Review*, and Dickens's journal, *All the Year Round*, carried a short article based on the ideas.[3] Mr Boffin's appropriation of the more notorious aspect of Müller's lectures is part of a wider concern in *Our Mutual Friend* with a whole spectrum of debates stemming from evolutionary thinking. During the months that Dickens was writing the novel, numerous reviews appeared of Lyell's *The Geological Evidences of the Antiquity of Man with Remarks on the Origin of Species by Variation* (1863), and in all of them reference is invariably made to the 'mounds' and 'dust-heaps' in Scandinavia and elsewhere that were being assiduously sifted for buried evidence of human origins. A review of Lyell's book in *All the Year Round* also comments on 'mounds' and 'refuse heaps'.[4]

The central question which underlies the diversity of artistic modes of *Our Mutual Friend* is that which arises from the anxiety, expressed by Matthew Arnold among others, as to the changing nature and state of health of British culture.[5] However, rather than locating fear for the survival of culture in the threat of anarchy, the novel is more Darwinian in its diagnosis of a degeneration of cultural vitality. Darwin's theories applied to nations suggested a pessimistic, wave-like movement of history in which primitive savagery progresses up the evolutionary scale towards the vigour and discipline of commercial and industrial civilization, until a certain point is reached when excess of prosperity and cultivation tilts the trajectory downwards as vitality is enfeebled by sensual indulgence. This was widely perceived to have been the fate of the Roman Empire, and journalism in the 1860s frequently drew warning comparisons between Britain's growing opulence and colonial expansionism and the decline of ancient Rome. In the text of *Our Mutual Friend*, Mr Boffin listens in horrified wonder to *The Rise and Fall of the Roman Empire*. The parts of Gibbon's narrative actually referred to are those in which the destruction of the Empire is hastened by the combined evils of savagery and venality.[6] It is as if the two extreme poles of the Darwinian evolutionary process meet in the death throes of the Empire. Prevalence of savagery and venality in the text of *Our Mutual Friend* would seem to hint forebodingly at a similar threat to British national and cultural vitality. Imagery of sterility, deserts and death abound, and characters in the text continually express desire to come back from the dead, or to find an opening out into a new world, such as Boffin hopes for in literature: 'This night, a literary man – *with* a wooden leg – [...] will begin to lead me a new life' (*OMF*, Book I, Chapter 5, p. 60). A similar link between literature and new life is made in the case of Charley Hexam, who is represented as a mixture of 'uncompleted savagery and uncompleted civilization' as he looks at the books in Veneerings's library with 'an awakened curiosity that went below the binding. No one who can read, ever looks at a book, even unopened on the shelf, like one who cannot' (*OMF*, Book I, Chapter 3, p. 28). Conventional wisdom would seem to imply here that education might complete the evolution towards civilization already half begun.[7] In this episode, Charley Hexam refers to the raising of Lazarus from the dead and to the escape of the Israelites from Egyptian slavery across the Red Sea. This latter image was a popular one in radical discourse throughout the nineteenth century for invoking the need to free labouring people from bondage to ignorance, poverty and injustice. However, references to the Nile and to Egypt had gained wider currency in the 1860s with Captain Speke's acclaimed discovery, in 1862, of the source of the Nile in East Africa. The importance of the floodwaters in annually bringing back life to the desert of Egypt became a commonly used trope in discussions of cultural regeneration.[8] In *Our Mutual Friend*, the Thames is frequently associated with baptism and renewal of life.

As these multiple, often oblique, textual references to Müller, evolutionary theory, cultural regeneration and degeneration indicate, the engagement of novelistic language with the most pressing issues of the 1860s can be parodic, figurative, fantastical, but rarely mimetically realist. It is largely this diversity of representational modes and social discourses available to it which allows the novel form to articulate such a complex sense of the material, intellectual, emotional, conscious and unconscious, serious and ephemeral currents that constitute its era; what Raymond Williams has described as a 'structure of feeling'.[9] If one were to predict what would be, or even argue for what should be, a major agenda for future nineteenth-century cultural and literary studies, it would be to increase our understanding of the complex and detailed engagements of literary texts with the social, cultural and economic conditions of their production. It has become clear that the large outlines offered by a Marxist analysis of social formations need to be much more finely contoured with specific, detailed understanding of the intermeshing of economic and cultural forces. Such contouring would include an informed mapping of the fluctuations, year-by-year almost and certainly decade-by-decade, in the shifting balances and negotiations of power within class fractions and between class blocks for hegemonic influence. Such fluctuations and shifts have to be recognized not only at the most obvious economic and cultural levels, but as equally and sometimes more powerfully enacted at affective and even visceral levels. Raymond Williams's term 'structure of feeling' seeks to understand cultural process in such a multi-dimensional way, and recently the work of Pierre Bourdieu has offered a methodological approach incorporating within a Marxist economic framework a critique of cultural practices ranging from food and sport to 'high' art. It may well be that the typical 'bagginess' of a nineteenth-century novel provides a particularly inclusive sense of the structure of feeling of Victorian culture at the moment it was written. This would be especially so of Dickens's novels with their intensely heteroglot language. My aim, in this essay, is to re-articulate the complex ways in which *Our Mutual Friend* expresses some of the main currents of feeling and social forces of the 1860s.

Marx's account in *Capital*, Volume III, of the transformation of 'the actual functioning capitalist into a mere manager, in charge of other people's capital' is itself graphically novelistic.[10] With the establishment of the stock market and joint stock companies by the 1860s, there arrives

> a new financial aristocracy, a new kind of parasite in the guise of company promoters, speculators and merely nominal directors; an entire system of swindling and cheating with respect to the promotion of companies, issue of shares and share dealings. It is private production unchecked by private ownership.
>
> Marx, *Capital*, III, 569

The fictional representation of the social world of Veneerings and Lammles bears striking resemblance to Marx's often satiric representation of speculative capitalism as a world dependent on ostentatious displays of opulence as a means of gaining fraudulent credit. However, while *Our Mutual Friend* demonstrates Dickens's imaginative grasp of the culture of share capitalism, the tone of the novel and its social satire seems closer to Marx's earlier *The Eighteenth Brumaire of Louis Bonaparte* (1852) with its emphasis on farce, parody, imaginary relations to reality and mediocrity posturing as the heroic.

In *The Eighteenth Brumaire* Marx looks closely at the interaction of culture with the economic forces of change. He explains how the bourgeoisie is inevitably forced to betray its own earlier political and cultural ideals. Its first historical role is heroic; it overthrows the old order based upon rank and privilege, establishing the dynamics of competitive capitalism which will bring about global transformation. However, once victory is secure, the bourgeoisie have

> a true insight into the fact that all the weapons which it had forged against feudalism turned their points against itself, that all the means of education which it had produced [...] all the so-called civil freedoms and organs of progress attacked and menaced its *class rule* at its social foundation and its political summit simultaneously.[11]

In order to safeguard its continued freedom to pursue economic ambitions and adventures, therefore, the bourgeoisie has to become an ever more repressive and reactionary cultural and political force. In particular, it has to curb all those cultural freedoms and innovations it previously championed in the arts, journalism, and public and political life. It becomes, says Marx, the Party of Order, representing any calls for reform as 'socialist', a threat to the family, property and religion. Under these proclaimed pieties it imposes a cultural order of absolute mediocrity; 'wherever it did not repress or react, it was stricken with incurable barrenness' (Marx, *The Eighteenth Brumaire*, p. 57). Thus, while in the early phase of its development, the bourgeoisie imaged its own epic struggles in the heroic myths of the Roman republic, its latter use of self-glorifying imagery is merely a veneer to mask its historical nullity. In place of the single-minded, ambitious men whose entrepreneurial energies in the early decades of the nineteenth century established a world capitalist order, 'a gang of shady characters push their way forward to the court, to the ministries, to the head of the administration [...] a crowd of the best of whom it must be said that no one knows whence he comes, a noisy, disreputable, rapacious boheme that crawls into braided coats with [...] grotesque dignity' (Marx, *The Eighteenth Brumaire*, p. 118).

Our Mutual Friend offers an identical representation of the political alliance of cultural reaction with disreputable financial dealing. The 'bran-new' man Veneering, 'sly, mysterious, filmy', and Alfred Lammle, 'a man of business as well as a capitalist' whose shady acquaintances are always 'coming and going across the Channel, on errands about the Bourse [...] all feverish, boastful, and indefinably loose', are perceived as commercial adventurers, living according to a culture of speculation, not just in shares on the world money markets, but in people (*OMF*, Book I, Chapter 2, p. 21; Book II, Chapter 4, p. 260). 'Speculate' is a much foregrounded word in the text.[12] This lawless financial adventurism explains the relationship to Podsnap. His solidity and repressive conventionality provides a sanctified veneer to film over the economic rapacity of commercial exploitation: 'It was a trait in Mr Podsnap's character (and in one form or another it will generally be seen to pervade the depths and shallows of Podsnappery), that he could not endure a hint of disparagement of any friend or acquaintance of his' (*OMF*, Book II, Chapter 4, p. 255). Podsnap as a representative man of the party of order establishes a stultifying mediocrity of culture, which blights the arts, morality and political life with barrenness while allowing a Darwinian struggle of each against each to prevail in financial dealings. It is a cultural and political order that fuses an encompassing dullness with competitive savagery. It is also represented as a social order that can image forth its own pretensions only as farce. As Wegg postures as a literary man, 'he compromised himself by no admission that his new engagement [...] involved the least element of the ridiculous. [...] His gravity was unusual, portentous, and immeasurable' (*OMF*, Book I, Chapter 5, p. 61). A similar portentous mediocrity characterizes Podsnap, 'executing a statuette of the Colossus at Rhodes' or gesturing in a farcical imitation of Napoleon as if his hearth rug were a world stage, 'See the conquering Podsnap comes, Sound the trumpets, beat the drums!' (*OMF*, Book I, Chapter 10, p. 119; Book I, Chapter 11, p. 137; Book II, Chapter 4, p. 255).

In *The Eighteenth Brumaire* Marx perceives culture as 'the sentiments, illusions, modes of thought and views of life' that structure individuals' imaginary relations to the actual social conditions of existence that they live (p. 39). In that sense, all those discourses and values of law and order, family and religion are merely the 'superstructure' (Marx, *The Eighteenth Brumaire*, p. 39) that sustains and disguises the capitalist class interest which requires economic freedom with absolute political stability. Their overthrow can only be effected, therefore, by changes at the basic level of relations of production. So, for the revolutionary destruction of the regime of cultural mediocrity, Marx relies upon the regenerative force of the proletariat. In particular, continuous capitalist expansion requires an intelligent and flexible work force 'available for the different kinds of labour required [...] the totally developed individual, for whom the different social functions are different modes of activity he takes up in turn' (Marx, *Capital*, I, 618). This necessitates educating the working class and thus capitalism's own expansionist logic will provide the proletariat with the weapons it needs for its ultimate triumph. Marx's optimism in respect of the revolutionary potential of working-class education seems not to have been vindicated.

Max Müller also expressed optimism in the regenerative role of the lower classes; their vigorous dialectical speech would re-energize a culture gone stagnant with over-cultivation and conventionality. Dickens would certainly have liked the way Müller also stressed the importance of metaphor in keeping alive creative thinking. In contrast to bow-bow theories of language which perceive an essential linkage between signifier and signified – words are imitation of natural sounds – it is the arbitrary relation of the two, Müller saw, which allows for the metaphorical dimension of language and it is this that makes new leaps of human thought possible, providing 'unlimited sway [...] in the formation of new ideas' (*Lectures*, II, 333). With this in mind, let's look a bit more closely at Noddy Boffin's 'bow-wow' outburst. It is, in effect, a contemptuous parody of John Harmon's confession of his feelings for Bella: 'if I had been so happy as to win her affections and possess her heart'. The implication of Boffin's response is that the words, so used, are well-worn hypocritical cant, which, through empty repetition, trip as readily from the tongue as the sounds 'bow-wow' and 'quack-quack' are mechanically attributed to dogs and ducks. However, the view of language implied in Boffin's usage is the opposite to that of the 'bow-wow' theory in that, far from positing an essential linkage between sound and meaning, Boffin is using the animal sounds metaphorically to suggest the duplicitous capacity of human speech – the frequent lack of connection between words and meaning. Boffin is represented, here, as having a shrewd insight into the language usage of many of his fellow characters.

Words in *Our Mutual Friend* are shown to be commodified in a system of surplus value.[13] Language may be said to have use-value in so far as it enables people to communicate information, feelings, ideas and values to each other. In a sense, words as use-value can be seen as a currency, like paper notes, symbolizing values guaranteed in substantiated human actions and attitudes, just as paper money needs a material gold reserve to guarantee its value. However, once forms of expression have accrued meanings and values through usage, they can be exploited to convey those meanings and values without the intentionality that previously guaranteed their worth; they gain surplus-value in effect. When Mr Venus asks Boffin, 'I have your word and honour?', Boffin retorts angrily, 'you have my word; and how you can have that, without my honour too, I don't know' (*OMF*, Book III, Chapter 14, p. 564). On this occasion, Boffin is naïve, since characters' words in the text more often than not exploit their surplus-value, drawing on the meaning they gained in previous usage. The majority of character discourse is constituted of inflated protestations of heroic and ennobling principles to mask inglorious, often sordid, motivation or farcical triviality. Wegg offers the most blatant example, appropriating the value created by real poets, he makes poetry 'tributary to friendship' and Boffin accepts the coin tendered as gratefully as if it were of sterling worth. It is the Veneering circle which the text most persistently represents as exploiting the language of mutuality and friendship for its surplus value. On the occasion of the Lammles' wedding, Veneering 'launches into a familiar oration', addressing the assembled guests as the 'oldest and dearest friends he has in the world' and concluding, says the narrator, with the time-honoured exhortation that with 'our hearts in our glasses, with tears in our eyes, with blessings on our lips, and in a general way with the profusion of gammon and spinach in our emotional larders, we should one and all drink to our dear friends the Lammles' (*OMF*, Book II, Chapter 16, p. 407).

Throughout the text, characters are represented as having well-stocked larders of commodified sentiments. When Mrs Lammle outlines to her husband her plan to defraud Mr Boffin, she knows exactly the form of bow-wows and quack-quacks required:

> 'Suppose my conscience should not allow me to keep to myself any longer what the upstart girl told me [...] Suppose I so repeated it to Mr Boffin, as to insinuate that my sensitive delicacy and honour – '
> 'Very good words, Sophronia.'
> *OMF*, Book III, Chapter 12, p. 547

What this indicates is that it is not just moral and social mediocrity that clothes itself in self-inflated portentousness, but its dark underside of corruption and adventurism also.

Paradoxically, though, characters are represented as having tenacious investment in their own rhetoric. They adhere doggedly to the bow-wow theory of language. Rogue Riderhood is represented as inducing in himself the strongest sense of outrage when his word is doubted. In this way, language constructs the imaginary relations in which characters are represented as living their real conditions of existence. Silas Wegg invents an intimate class relationship to 'Our House', aggrandizing himself in his imagination with delusions of being on a footing of confidential trust with 'the Quality' (*OMF*, Book I, Chapter 5, p. 54). The arts, too, are brought into service of imaginary social relations, literature, painting, sculpture and music being conscripted to provide banal and expurgated representations of bourgeois life. Podsnappery disallows any alternative version of reality (the starvation of the poor, for instance) to those mutually self-mirroring illusions of national and individual glory that feed egoism and complacency. This is a bow-wow realist theory of art, asserting a total identification between representation and life, and allowing no freedom for the inventive, visionary capacity of metaphor. These two apparently contradictory usages of language consisting of a practice of fantasy and duplicity with an insistence on the absolute bonding of signifier to signified are the linguistic equivalent of the duality of rapacious economic free-for-all with moral and cultural repression. The disease of language is shown as indissolubly linked to the financial venality of economic Darwinism.

On the one hand, then, what is represented in *Our Mutual Friend* is a culture which insists that words are bonded to meaning and value, and on the other, one where words have become an inflated currency adrift from any guarantee of intentionality. The assertion of belief is the necessary condition of possibility for the practice of duplicity. This degeneration of language presents a crisis for any writer who aims to envisage and promote a new cultural order. Where can a culture which has devalued its own expressive currency find the means of self-renewal? The novel's ending seems to follow Müller in suggesting a revitalization of imaginative energies through an alliance of cultivation (in the form of the patrician disaffection of Wrayburn and his linguistic playfulness) with the 'dialect-speaking classes'. It is certainly those characters who are poor who articulate desire for transformation. Jenny Wren's self-naming enacts an imaginative transformation of her crippled body and degraded life; she metaphorically assumes an identity of bird-like lightness. Human beings, it is thus suggested, often need some form of imaginary relation to their actual conditions of existence and this capacity to transform a sordid or crippling reality is clearly linked to any potential human beings have to renew their culture. During the evolutionary debates, it was this capacity for ennobling spiritual desires that was held to elevate the human from the brute. However, unlike Müller and Marx, *Our Mutual Friend* does not represent the poor as untouched by the culture of commodity capitalism. The text frequently suggests that their desire for transformation finds its most readily available outlet in a commodified taste. Mr and Mrs Boffin, having drudged all their lives in 'one dull enclosure', discover an inexhaustible source of pleasure in 'the variety and fancy and beauty of the display in the [shop] windows, [...] as if the principle streets were a great Theatre' (*OMF*, Book III, Chapter 5, p. 461). The wife of the Reverend Frank Milvey is represented as having had to repress 'many pretty tastes and bright fancies' (*OMF*, Book I, Chapter 9, p. 109). The emphasis upon taste and its linkage with fancy or imagination, as in this passage, is new in Dickens's work. Its most interesting occurrence is when Jenny Wren says 'we Professors who live upon our taste and invention, are obliged to keep our eyes always open' (*OMF*, Book IV, Chapter 9, p. 714). She is explaining how even at the funeral of her father she saw the potential of transforming the officiating clergyman into a doll who would perform at dolls' weddings not burials. The use of the word 'taste' in these contexts associates it with an inherent desire for transcendence; it becomes evidence of the impulse to escape all that brutishly holds down the imaginative aspiration to a better existence.

In contrast to this usage, the text also represents taste as an indissoluble bonding of signifier and signified. Characters are defined by their taste. Style is the man – Podsnap's plate and furniture characterized by 'hideous solidity'; it is also the woman – Lady Tippins of whom 'you could easily buy all you see of her in Bond Street', and who reckons worth in purely consumer terms: 'Bride [...] thirty shillings a yard [...]. Bridesmaids [...] twelve and sixpence a yard' (*OMF*, Book I, Chapter 11, p. 135; Book I, Chapter 10, pp. 122–3). It is helpful to understand this linkage of taste with transcendent imagination on the one hand and with commodified identity on other, by relating it to Pierre Bourdieu's argument that taste is 'one of the most vital stakes in the struggles fought in the field of the dominant class'.[14] In a commodity culture, he claims, a class is defined as much by its relations to consumption as to production (*Distinction*, p. 483). In particular, taste is the site of a struggle for power between different fractions of the dominant class, between those whose status depends largely upon economic capital and those whose claims to status rests upon cultural capital. Cultural capital is largely inherited, bequeathed by the traditions and upbringing offered in families that have long enjoyed cultural dominance, and consolidated by educational capital gained at the most eminent public schools. Bourdieu argues that taste functions as the prime means of social classification by perceiving goods and lifestyle as materialized symbols of a person's distance from physical necessity. This is symbolized most obviously by conspicuous displays of luxury, such as Veneering's sumptuous dinners which function as an index of wealth, far removed from the need of food to live. However, those without economic capital use their cultural capital to assert status through claims to aesthetic and spiritual distinction which, because the style is acquired from family upbringing, can assert itself as natural, inborn superiority, despising the vulgar showiness of the *nouveau riche*. It styles itself as discernment or taste which cannot be bought. Bourdieu writes, 'It confers the self-certainty [...] of possessing cultural legitimacy, and the ease [...] which bourgeois families hand down to their offspring as if it were an heirloom' (*Distinction*, p. 66). Cultural capital allows distance from necessity to be symbolized in a personal and bodily style of detachment, nonchalance and self-possession. Intra-class struggles fought out at the level of symbolization around issues of taste and cultivation become particularly intense at times of sudden increases in new wealth or when a traditionally eminent group feels itself threatened. The 1860s were one such moment when, for the first time, commercial wealth began to outstrip wealth based on landed property.[15] In addition, it was a time when the non-commercial middle-classes felt threatened by the imposition of competitive entry into the civil service and by renewed demands for equality from the organized working class.[16]

Taste is relational; what is perceived as 'high' and 'refined' can only be so in opposition to something else perceived as 'low' or 'coarse'. Bourdieu argues that an oppositional system of symbolic terms is brought into play in the struggle between economic capital and cultural capital, and that underlying its use is a persistent reference back to the notion of freedom from physical necessity. Thus earnestness is opposed to playfulness, heaviness to lightness, pedantry and pomposity to wit, coarseness to refinement, and most importantly of all, materialism to spirituality. An ideology of 'inborn taste' converts material cultural objects and practices into signifiers of natural nobility of soul. To win in this game of culture, Bourdieu claims, one must acquire the ability to play it playfully with an assumption of carelessness (*Distinction*, pp. 71, 330). This is strikingly the pattern of oppositions that constructs the characters of Podsnap and Wrayburn in *Our Mutual Friend*. While the word 'heavy' is used frequently in connection with Podsnap, the word 'light' is constantly associated with Wrayburn to suggest his physical ease of manner ('so light and careless') and his playful disposition, disdainful of all vulgar earnestness (*OMF*, Book II, Chapter 2, p. 235). Bourdieu goes on to suggest that whereas those depending on economic capital will expect from art 'a reinforcement of [...] self-assurance' in realist representational forms, the intellectual fractions of the bourgeoisie want an avant-garde art that will challenge materialism and provide another means of symbolizing their cultural and aesthetic capital (*Distinction*, p. 293). They want art forms that state their distance from necessity by a rejection of realism and by an aesthetic detachment and playfulness. We have here the separation of the arts according to Podsnap or the arts according to Wrayburn.

Clearly, for any writer who believes that literature should contribute to cultural elevation, there is seductive power in the ideology of cultural capitalism, stylized as disinterestedness, inborn taste, 'sweetness and light', as Arnold terms it.[17] However, recognizing that this opposition of styles and tastes is the symbolic ground of struggle for dominance within the dominant class itself suggests that there are strict limits to the transformative potential of Wrayburn's form of social disaffection. It is style adrift from intentionality. This becomes strikingly clear if we look at the representation of the inter-class confrontation between Wrayburn and Headstone in the novel in terms of Bourdieu's pattern of cultural oppositions. Headstone bases his claim to respect and status on his newly acquired educational capital. This is the traditional route to new conditions of existence aspired to by many of the nineteenth-century working class; it is also the means that Marx looked to for a revolution of the means of production and hence of culture. Whatever visions of hopeful transcendence inspired the origins of Headstone's journey out of necessity through education, the trajectory was likely to be a brutally foreshortened one. All the debates around the proper education for the working class and especially those concerning pupil-teacher education were intent on drawing and maintaining distinctions. The revised code ensured that working-class children becoming teachers gained no knowledge which might allow them entry into the 'cultivated' classes. Whereas public-school education conferred cultural legitimacy, the working of the revised code, through its system of payment by results, meant that the teacher was perpetually on trial, continually forced to prove his or her meagre competence.[18] The more Headstone boasts of certificates, the more hopelessly he loses ground in the 'culture-competition', betraying the inevitable narrowness of his view of knowledge.

Bourdieu compares the emphasis on sport in public-school education with the 'barrack-like' regime of working-class education (*Distinction*, p. 93). Whereas the former imparts bodily ease and well-being, the latter engenders a physical awkwardness. Class, he writes, is encoded in the 'most insignificant techniques of the body – the way of walking or blowing one's nose, ways of eating or talking' (Bourdieu, *Distinction*, p. 466). He continues, 'One's relationship to the social world and to one's proper place in it is never more clearly expressed than [...] in the space one claims with one's body in physical space, through a bearing and gestures that are self-assured or reserved, expansive or constricted' (*Distinction*, p. 474). The representation of Headstone and Wrayburn is structured upon precisely this opposition between bodily ease and disease. Wrayburn, as seen by Headstone for the first time, is described as 'coolly sauntering [... with] a certain lazily arrogant air [...] holding possession of twice as much pavement as another would have claimed' (*OMF*, Book II, Chapter 1, p. 229). Every description of Headstone makes reference to his physical constraint and discomfiture. In particular, he is shown as having no control over his body's physical manifestations: his face burns red and turns deathly white, his hands clutch convulsively, above all, he continually sweats (*OMF*, Book II, Chapter 11, pp. 339–40). Headstone has gained certificates but his body stays relentlessly lower class. His eventual succumbing to seizures or fits, the most visually distressing loss of bodily control, seems to symbolize an inherent physical degeneration. Nothing could make clearer than this representation the structuring opposition between bodily necessity and distance from that necessity which generates and upholds notions of cultivation and taste. The confrontation between the two men is a confrontation between oppositional values: those based upon the common needs of bodily life and those that claim transcendence from those needs, in disembodied spiritual distinction.

The hatred between these two characters, figuring wholly antagonistic value systems, is conveyed in writing charged with visceral revulsion. Lizzie Hexam functions in the text merely as an index to measure the degree of physical desire and disgust each man evokes. In *Distinction*, Bourdieu writes of 'the "looking-glass self" reflected in the reactions of others' and of the way those legitimated by upbringing and education may experience their physical being as a 'vessel of grace' (p. 207). This is how Lizzie Hexam is represented as perceiving Wrayburn in his contrast to Headstone: 'going on at her side, so gaily, [...] so superior in his sallies and self-possession to the gloomy constraint of her suitor [...] that his lightest touch, his lightest look, his very presence beside her in the dark common street, were like glimpses of an enchanted world' (*OMF*, Book II, Chapter 15, p. 399). Headstone's 'looking-glass self' comes from the class contempt imaged in Wrayburn's face. This look of 'cold disdain of him, as a creature of no worth' crushes Headstone into the dirt; enforcing upon him recognition that he cannot escape necessity (*OMF*, Book II, Chapter 6, p. 285). Despite certificates there is no transcendent evolution for him out of confining physical being into a light and careless refinement. The narrative closes him into ever narrower limits, rendering his final night in terms of physical extinction, 'the dark lines deepening on his face, [...] its surface turning whiter and whiter as if it were being overspread with ashes, and the very texture and colour of his hair degenerating' (*OMF*, Book IV, Chapter 15, p. 779). All that is left of his trajectory is for him to cast himself back into primal slime and mud.

In the Darwinian struggle for survival, sexual dominance is the key to future evolutionary forms. The imaginary solution the narrative offers for the malaise of language and culture it so powerfully analyses is strikingly like that of Müller: a sexual union of cultural capital with the dialect-speaking classes. The grouping of characters brought together at the end of the novel – the Boffins, the Harmons, the Wrayburns, Jenny Wren and Sloppy – represents a utopian community: lower-class trueness of heart, responsibly administered wealth, cultural refinement and working-class artistry and craftsmanship in the form of Jenny Wren's dolls and Sloppy's carpentry. The whole community is envisaged far from the 'common streets', safely enclosed in 'Our House', tastefully refurbished, full of flowers and birds – the epitome of 'sweetness and light'. The force which has been rigorously excluded from this regenerative cultural mix is the figure of the educated working class, ruthlessly cast back onto brutish origins.

Within the textual diversity of *Our Mutual Friend*, the use of parody and grotesquerie embodies a critique of financial and verbal venality with an appositeness which could not easily be matched by realist form. Given the linkage of a bow-wow version of realism with cultural corruption, it is not surprising that the text eschews mimetic form. However, the imaginary solution the narrative works towards degenerates in artistic vitality the more entangled it becomes in the ideology of distinction. For those in the 1860s appalled by aggressive commercial Darwinism on the one hand, and on the other repelled by the physical energies they felt pushing up from below, an ideological fiction of transcendence built a tasteful mansion of inborn, apparently classless refinement. *Our Mutual Friend* registers in visceral terms where the lines of distinction and exclusion in these intra- and inter-class struggles were most deeply cut. Paradoxically, though, the full anguish of such exclusion is most powerfully driven home by writing which moves closest to realist mode. In a novel full of people represented as shams, pretenders and mistaken identities, only Headstone takes on fictional substance; he is the only character the writing endows with imaginative force. Headstone is represented as saying to Lizzie Hexam, 'if it is any claim on you to be in earnest, I am in thorough earnest, dreadful earnest' (*OMF*, Book II, Chapter 15, p. 390). That earnest claim from the aspiring working class was denied in the careful class limits set to the cultural horizons offered in the various education provisions made in the last decades of the nineteenth century. Looking at what has been on offer in the twentieth century, we might have to recognize Dickens's prescience when he so grimly seals Headstone off from the possibility of a regenerative future. Of the two opposing systems of value – the one based upon the common necessities of life and the other upon claims to transcend those needs – it is the latter which still organizes our educational hierarchies and, frequently, our aesthetic judgements.

Notes

1. *Our Mutual Friend* (Harmondsworth: Penguin, 1997), p. 583 (Book III, Chapter 15). Further references will be given in the text.
2. Max Müller, *Lectures on the Science of Language,* ed. Roy Harris, 2 vols (London: Routledge/Thoemmes Press, 1994), I, 344. Further references will be given in the text.
3. *Edinburgh Review*, 115 (1862), 67–103; 'Picture Words', *All the Year Round*, 8 (1862), 13–16.
4. 'How Old Are We?', *All the Year Round*, 9 (1863), 32–7.
5. Matthew Arnold, *Culture and Anarchy* (1868) (Cambridge: Cambridge University Press, 1960).
6. See Edward Gibbon, *The Decline and Fall of the Roman Empire* (Harmondsworth: Penguin, 1981), pp. 113–32.
7. Jeremy Tambling, *Dickens, Violence and the Modern State: Dreams of the Scaffold* (Basingstoke: Macmillan, 1995), points to the political implications of Charley Hexam's evolution (p. 215).
8. See, for example, a leading article celebrating the Nile as 'fountain of life' in *The Times*, 29 November 1862. I discuss this pattern of imagery in more detail in *Dickens's Class Consciousness* (Basingstoke: Macmillan, 1991), pp. 120–1.

9. Raymond Williams, *The Long Revolution* (Harmondsworth: Penguin, 1965), p. 64.
10. *Capital*, 3 vols (Harmondsworth, Penguin, 1991), III, 56. Further references to all three volumes will be given in the text.
11. *The Eighteenth Brumaire of Louis Bonaparte*, rev. edn (London: Lawrence and Wishart, 1954), p. 56. Further references will be given in the text.
12. See *OMF*, Book I, Chapter 5, p. 54; Book I, Chapter 10, p. 129; Book II, Chapter 8, p. 317; Book II, Chapter 12, p. 345; Book II, Chapter 16, p. 409; Book III, Chapter 15, p. 578.
13. According to Marx, a certain number of hours of human labour (say 10) result in a product (a coat) which has use-value (warmth). Thus human activity/labour produces use-value. By analogy, we may say that so much collective human labour (as speech linked to action) results in the production of shared meanings which have use-value. The capitalist system transforms products/use-value into commodities bearing surplus-value. If the worker can be made to produce two coats in ten hours he is paid the same hourly rate and the capitalist gains the extra value as surplus. Again by analogy, the use-value of words becomes commodified into surplus value when meaning produced by previous human activity (speech/actions) is appropriated and exploited in new 'unearned' usage. See *Capital*, I, especially 293–306.
14. *Distinction: a Social Critique of the Judgement of Taste*, trans. Richard Nice (London: Routledge, 1986), p. 11. Further references will be given in the text.
15. For an account of this, see Harold Perkin, *The Rise of Professional Society in England since 1880* (London: Routledge, 1989), p. 64.
16. Reform of the civil service began in 1853; although competitive entry was not established until 1870, the ethos of merit was adopted well before this date. See Harold Perkin, 'The Professionalization of Government', in *The Origins of Modern English Society, 1780–1880* (London: Routledge and Kegan Paul, 1969), pp. 319–39.
17. Arnold, *Culture and Anarchy*, p. 54.
18. The Taunton Commission on the Public Schools (1868) defended the exclusive teaching of a classical education rather than sciences and modern languages on the grounds of social status and values. Members of the higher classes without wealth 'had nothing to look to but education to keep their sons on a high social level. And they would not wish to have what might be more readily converted into money, if in any degree it tended to let their children sink in the social scale' – quoted in Martin J. Wiener, *English Culture and the Decline of the Industrial Spirit, 1850–1980* (Cambridge: Cambridge University Press, 1982), p. 18. For a discussion of the Revised Code of Education (1863) which instituted a system of payment by results, see Phillip Collins, *Dickens and Education* (London: Macmillan, 1963), pp. 159–71.

13
'The Mothers of our Mothers': Ghostly Strategies in Women's Writing

Anthea Trodd

Anne Thackeray Ritchie's *A Book of Sybils* (1883) is an early attempt to construct a women's tradition in writing, and thus seeks ways of addressing such issues as how to situate women within the historical narrative of the literary tradition, what specific connections to establish between them, what value to place upon sometimes exiguous traces.[1] The book reprinted a series of essays from the *Cornhill Magazine* on early-nineteenth-century women writers, the literary grandmothers of Ritchie's generation, and explored ways in which their lives might be recreated, an exploration in which Alice Meynell was Ritchie's principal successor, and Virginia Woolf the ultimate legatee.[2] Ritchie's attempt to construct a tradition is extremely tentative, and full of apologies for the comparative insignificance of most of the women's writing of the period she describes, and for her difficulties in reconstructing their lives and work. The problem to which she continually returns is how little of their significance can be recovered. Carolyn Steedman has argued that this has been the central recognition of researchers in women's history during the last few decades:

> One legacy is an altered sense of the historical meaning and importance of female *insignificance*. The absence of women from conventional historical accounts, discussion of this absence (and of the real archival difficulties that lie in the way of presenting their lives in a historical context) are at the same time a massive assertion of the

littleness of what lies hidden. A sense of that which is lost, never to be recovered completely, is one of the most powerful organizing devices of modern women's history.[3]

Ritchie's particular anticipation of this recognition of irretrievability organizes her tentative tradition of women's writing. Her narrative begins with a ghost story, and invokes the supernatural throughout as a means to describe the occult presence of women in literary history, the difficulty of evaluating its traces, and her worries about women's relation to the official literary tradition. As a minor genre of low critical reputation associated with grandmothers and their old wives' tales, the ghost story also serves as a useful disclaimer of any pretensions in the exercise. Ritchie's method of justifying her attention to a female tradition by invoking the supernatural is paralleled in the work of other women writers of the late nineteenth century, and this essay will argue that such reference was expressive of the difficulties of women looking in history for the presence of women.

Although she wrote several well-received novels in her earlier career, such as *The Village on the Cliff* (1867) and *Old Kensington* (1873), Ritchie was increasingly associated with the recreation of the lives of dead writers. It was this role which her step-niece, Virginia Woolf, celebrated in her obituary in the *Times Literary Supplement* in 1919. 'She will be the unacknowledged source of much that remains in men's minds about the Victorian age. She will be the transparent medium through which we behold the dead'.[4] Ritchie memorialized her father in the prefaces for the Biographical Edition of his works, and evoked other great Victorians in *Records of Tennyson, Ruskin and Robert and Elizabeth Browning* (1892), *Chapters from some Memoirs* (1894) and *From the Porch* (1913). *A Book of Sibyls* was her most distinctive work. Robert Louis Stevenson, the rising star of the new genre of masculine romance, complimented Ritchie on her evocation of a forgotten past:

> The faces and the forms of yore,
>   Again recall, again recast;
> Let your fine fingers raise once more
>   The curtains of the quiet past;
> And there, beside the English fires
>   That sung and sparkled long ago,
> The sires of your departed sires,
>   The mothers of our mothers show.[5]

Stevenson's tribute endorses both the book's claim to have conducted an investigation of a quasi-supernatural kind, and its assumption that the skills required are essentially female and domestic.

The apology which begins *A Book of Sybils* is couched in the form of a ghost story. In a Hampstead street, wandering in 'a pleasant confusion of past and present' induced by personal and literary memories, Ritchie suddenly glimpses a woman in anachronistic costume whom she recognizes as Anna Barbauld, accompanied by her husband. It seems that communication between the dead and living writers is about to take place, a potential meeting of sympathies. 'The lady came forward, looking amused by my scrutiny, somewhat shy I thought – was she going to speak? And by the same token it seemed to me that the gentleman was about to interrupt her' (*Sybils*, p. 5). At this point, however, Ritchie's 'young companion', oblivious of the ghostly encounter, intervenes, and the ghosts vanish. There is, however, an aftermath. 'Many well-authenticated ghost stories describe the apparition of bygone persons, and lo! when the figure vanishes, a letter is left behind!' (*Sybils*, p. 6). In this case a packet of Barbauld's letters is coincidentally left at Ritchie's home later in the day, on the basis of which, and the sympathies evoked by the ghostly encounter, she begins her research for the essay which will reconstruct Barbauld.

An attempt to reconstruct a tradition of women's writing thus begins as a ghost story, the features of which include an involuntary passage into the past which justifies the reconstruction, potential communication between the two women which is, however, interrupted on both sides of the time divide, and finally an attempt to re-establish through research the lost communication. Although Barbauld's apparition materializes to provoke the exploration of her work, no great claims are made for its importance in the mainstream literary tradition. The emphasis is on what she might have achieved in happier circumstances:

> if the daily stress of anxiety and perplexity which haunted her home had been removed – difficulties and anxieties which may well have absorbed all the spare energy and interest that under happier circumstances might have added to the treasury of English literature.
>
> *Sybils*, p. 22

The essay develops an opposition between the male writing in 'the treasury', and the scattered writings outside for the study of which other justifications must be found. At the conclusion Ritchie is even more explicit about the kind of niche Barbauld and her contemporaries occupy in the literary tradition: 'it is certain that the modest performances of the ladies of Mrs Barbauld's time would scarcely meet with the attention now, which they then received' (*Sybils*, p. 50). Barbauld's importance lies elsewhere, in her children's hymns, and in the 'Ode to Life' (1825), which, 'like sweet church bells on a summer evening' (*Sybils*, p. 47), comforted the dying. Barbauld's work thus affords a kind of cradle-to-grave protection, offering support at the most vulnerable stages of life, a claim reinforced by Ritchie's reference to her own grandmother:

> the hand that patiently pointed out to her [Ritchie], one by one, the syllables of Mrs Barbauld's hymns for children, that tended our childhood, as it had tended our father's, marked those verses ['Ode to Life'] one night, when it blessed us for the last time.
>
> *Sybils*, p. 48

In this reference the functions of Barbauld and of the grandmother are merged, and Barbauld's career as a writer is seen as flourishing in a domestic sphere quite distinct from the mainstream tradition. What her 'modest performances' offer is an opportunity to recover contact with a lost world of strong feeling; 'the stock of true feeling, of real poetry, is not increased by the increased volubility of our pens' (*Sybils*, p. 50). Indeed it appears that the very modesty and limitation of Barbauld's poems allow better communication with this past world.

Ritchie accommodates the subject of her third essay, Amelia Opie, to precisely the same model of woman writer. There is the explicit disclaimer that the works are significant within the mainstream tradition. 'It would be impossible to attempt a serious critique of Mrs Opie's stories. They are artless, graceful, written with an innocent good faith which disarms criticism' (*Sybils*, p. 184). Much of the essay searches in other evidence – portraits, letters, the comments of contemporaries – for fleeting traces of Opie, evoked in impressionistic tableaux as 'a young and charming figure, with all the spires of Norwich as a background, and the sound of its bells' (*Sybils*, p. 153). Again it is precisely the inadequacy of the published evidence which is the clue to the nature of women's lives in this lost world.

It is this emphasis on the irretrievable which explains the title, chosen shortly before publication, and apparently suggestive of writers whose prophetic power is conveyed in fragmentary form only to succeeding generations.[6] There is, however, no attempt to ascribe extraordinary prescience to the four writers treated, or to suggest that their surviving works are only fragments of their written oeuvre. The idea of sibylline fragments, sustained throughout the book, refers not to the incomplete survival of the subjects' work, but to the incompleteness of their achievement; their annalist must assess what in other circumstances they might have created. The main force of supernatural suggestiveness in 'sybil' is thus displaced from the writers described towards the writer who reconstructs. The opening ghost story establishes the difficulties of raising the dead from the obscure past, and the mixture of supernatural good luck, perception and sympathy needed. The transference of the word 'sybil' to the author who reconstructs is reinforced by the dedication to Mrs Oliphant, 'dear Sybil of our own'. Woolf's 1909 satirical sketch on Victorian literary women, 'Memoirs of a Novelist', comments in passing on their habit of applying the term 'sibyl' to each other, suggesting the term's currency for women writers seeking to describe themselves.[7]

Ritchie's strategy of organizing her literary history around ideas of the occult and irretrievable reaches its height in the enormously influential essay on Austen, which was the first written and concludes the book.[8] Austen is described as 'writing in secret, putting away her work when visitors come in, unconscious, modest, hidden at home in heart, as she was in her sweet; womanly life' (*Sybils*, p. 199). In this essay the narrative of the sibylline fragments is sustained in the description of Austen's secretive, interrupted mode of writing, famously extended by Woolf in *A Room of One's Own* (1929), and the suggestion that in other circumstances more work might have been written links Austen to Barbauld and Opie. Ritchie's version of Austen dwells on her as aunt, and naturalizes her within a secluded country setting through the metaphor of a nestbuilding songbird:

> She built her nest, did this good woman, happily weaving it out of shreds and ends, and scraps of daily duty, patiently put together; and it was from this nest that she sang the song, bright and brilliant, with quaint thrills and unexpected cadences, that reaches us even here through near a century. The lesson her life seems to teach us is this: Don't let us despise our nests.
>
> *Sybils*, pp. 226–7

Austen is thus accommodated within the women's occult tradition Ritchie is establishing, despite the brilliant survival of her works within the treasury. Her writing becomes an extension of ordinary household routine, and is then further naturalized by the bird metaphor, which suggests it is a spontaneous accompaniment to domestic activity. At the same time the reader's distance from Austen is exaggerated, as sixty odd years become 'near a century', the effect enhancing both the difficulty of reconstructing this lost world, and the urgent need to resume contact with its right feeling. The novels, which appear throughout the essay almost entirely in references to their characters, or 'people', are subsumed in Austen's life. Their real value is to connect us with 'what we have lost in calm, in happiness, in tranquillity' (*Sybils*, p. 209). In evoking a lost world of strong feeling, their value is of the same kind as the apparition of Barbauld.

In the justification Ritchie offered for a series on women writers, she created the outline for a tradition of women's writing, in which the work's comparative obscurity, and lack of objective importance, becomes a major part of its examinable significance, and the primary objective is to recover a lost world of strong feeling, the memory of which has been nearly obliterated through the comparative absence of women's writing from the canon. In this outline Ritchie could afford to be sweeping in her external evaluation of her subjects' achievements, to concede that they were 'rather literary women than actual creators of literature' (*Sybils*, p. 152). They had, after all, suffered from the kind of interruption which overtakes the apparition of Barbauld as she prepares to speak to Ritchie. The role of the writer who recreates the lives of these women is to look beyond the mostly inadequate traces to speak what they had not spoken themselves.

One distinctive problem is presented by the writer whose traces are not occult, who has accommodated herself within the mainstream tradition. The second essay, on Maria Edgeworth, is the longest, and conspicuously the least interested in its subject. Ritchie acknowledges that the extensiveness and achievement of Edgeworth's work is inappropriate to the quasi-supernatural process of recreation in which *A Book of Sybils* is engaged; Edgeworth's ghost cannot walk for Ritchie. There is no potential contact of sympathy: 'some indefinite sentiment tells people where they amalgamate and with whom they are intellectually akin' (*Sybils*, p. 124). Ritchie is conscious that 'she has scarcely done justice to very much in Miss Edgeworth, which so many of the foremost men of her day could appreciate' (*Sybils*, p. 124). Edgeworth belongs within the mainstream literary tradition. There is little to be done for her by the sibylline interpreter of scattered fragments and occult traces, despite some attempts to accommodate her to an approximation of the model by dwelling especially on her writing for children, and taking the most extensive quotation from her account of her meeting with another woman writer, Mme de Genlis. Edgeworth belongs to another, public, tradition, like Eliot, on whom Ritchie declined to write on her death in 1880.[9]

The difficulty of dealing with Edgeworth is a consequence of Ritchie's central emphasis on the near irretrievability of women's history. Bonnie Smith, in discussing women's historiography, has argued that for both nineteenth- and twentieth-century women historians:

> oppression was the principal theme of the history of woman; her historic visibility either undermined that theme or testified to her lack of feminist or other virtues. Under these circumstances the old question – which women actually merited historical treatment – remained germane.[10]

Ritchie's ghost stories ask that question as they give priority to the fugitive presence, the barely glimpsed and inarticulate figure, the interrupted contact, signs of the largely submerged history of women. Ritchie is explicitly happier with Barbauld and Opie, who said so little of what they might have said because they were always being interrupted, than with Edgeworth, whose writings were too extensive and admired to be described as submerged. The difficulty of dealing with Austen organizes her influential essay; Austen remains essentially private, hidden, unknown, and therefore, despite her high visibility as writer, paradoxically can be the patron saint of the woman submerged and unknowable in history.

For other late-Victorian women writers the ghost story was a way of talking about women's forgotten history. In her most famous ghost story, *A Beleaguered City* (1880), the sibylline dedicatee of *A Book of Sybils*, Margaret Oliphant, also describes a half-forgotten female culture, the value of which can be recognized only through a magical recreation of the past. In this novella the sun fails to rise one day on a small town in the Haute Bourgogne, where gender segregation has reached an extreme point; almost all the men are rationalists, and religion is seen, in an often repeated phrase, as 'an affair of the grandmothers'.[11] In the darkness an invisible army of the dead take over the town, evicting the living by sheer pressure of numbers. During the eviction a number of the women, but very few of the men (and the materialistic narrator is not among them), recognize lost relatives amid the army of the dead. In particular the narrator's wife recognizes their lost child. Unmanned by invisible terrors, the leading citizens regress briefly to a state of childhood, until daylight returns to the city and they are able to re-enter. The lost world of the grandmother, the world of church bells and strong ties with the dead, has been briefly evoked by a supernatural act, for the duration of which the male public world has been forced to re-establish continuity with the past and recognize the value of the alternative female world of feeling. (Kipling's 1904 ghost story, 'They', reworking the story of *A Beleaguered City*, acknowledges the power of its representation of women's experience as a ghost story which underlies the male materialist world. The male narrator stumbles into an alternative world of ancient tranquillity and dead children, and recognizes that as an adult male he must reject this alluring lost world of the past in which it is permissible for women to live.)

Oliphant's work is a formal ghost story; two works by women academics, both influential though widely divergent in kind, also turn to ghost stories to help them talk about the need for a history to record the under-recorded. The 1890s work of Jane Harrison, the first important English woman academic classicist, was concerned with recovering the lost antecedents of the Hellenic world known in literature. Robert Ackerman and Hugh Lloyd-Jones have both argued that Harrison's exaggerated diffidence about her linguistic abilities led her to redirect her research away from the study of Greek literature towards the less prestigious study of artefacts and practices.[12] In an academic culture where mastery of classical literature was long-established as a crown of male scholarship, Harrison doubted her ability to hold her own, and turned from the literary record to a less explored world, a direction confirmed by a trip to Greece, in which actual contact with places and artefacts encouraged her belief in her ability to recreate a world of feeling and belief missing, or imperfectly adumbrated, in the literature. In her first major book, *Prolegomena to the Study of Greek Religion* (1903), Harrison explores the forgotten ghost-stories and ghost-aversion rituals which, she argues, are the substratum of fully known Greek culture, and from which Greek drama and religion developed. This exploration is associated in Chapter 6, 'The Making of a Goddess', with the occult history of women. Harrison attempts to trace the evolution of the Olympian goddesses, but finds that the traces of the process, still highly recoverable for male deities, are largely obscured for female. It is possible only to note the general degeneration of the female deity from the 'early matriarchal, husbandless goddesses' who were patrons of the heroes. 'With the coming of patriarchal conditions this high companionship ends. The women goddesses are sequestered to a servile domesticity, they become abject and amorous'.[13] There is then a suppressed history of women in this evolution from ghost to goddess; as ghosts evolve towards the Olympian deities of literature the process by which women lost their status is obscured. They appear in the historical record only as barely traceable ghosts, or as the literary concepts of the poets.

Although Harrison's narrative depends on current discussions of early matriarchal societies, her admiration for this lost history is tempered by her concern that this early female power had rested on a false belief in women's magical powers:

> Matriarchy gave to women a false because a magical prestige. With patriarchy came inevitably the facing of a real fact, the fact of the greater natural weakness of women. Man the stronger, when he outgrew his belief in the magical potency of women, proceeded by a pardonable practical logic to despise and enslave her as the weaker.
>
> *Prolegomena*, p. 285

Harrison's dilemma is clear. She is strongly attracted to the idea of recovering a lost world in which women enjoyed power and high esteem, in which the feelings which now define them as inferior were part of that esteem. She is partisan for the neglected history which lies behind the completed grandeur of Greek art and literature, the history of ghosts and ghost-aversion rituals, and angry that the emotions and beliefs which created and sustained these rituals, and, above them, the whole edifice of Greek art, the emotions and beliefs of ordinary people, women, the uneducated, have been unexamined. The history of women belongs to that occult world. At the same time, as an academic and rationalist, who lost her faith early after reading *The Origin of Species* (1859), she is conscious that a prestige based on magic, the only great prestige historically available to women, cannot be revived. Her fighting comments on the decline of the prestigious deities of matriarchal society are rapidly undercut by the admission that a prestige based on false pretensions to magic needs to disappear. (Her autobiography recalls her friend Ritchie with an indulgent disapproval: 'she never, I think, had her delicate feet quite on the ground'.[14])

The other academic example of female recourse to the ghost story is of a very different kind. In 1911 Charlotte Moberley and Eleanor Jourdain published *An Adventure*, the record of their alleged experience at Versailles in 1901 of time-travel to 1789, which became one of the most widely discussed psychic events of the earlier twentieth century.[15] Although they also belonged to the first generation of women academics, they were not distinguished scholars like Harrison, but primarily administrators. According to her ex-student and executor, the medieval historian Joan Evans, Moberley's 'ideas of historical research were largely derived from her godmother, Charlotte Yonge'.[16] If Harrison felt diffidence about her scholarship, Moberley and Jourdain had every reason to feel greater diffidence. *An Adventure* might be seen as their attempt to explore their academic discipline, and the historical situation of women. After detailing their psychic experience, and ten years of research in French historical archives, which have convinced them that experience offered them an authentic short cut to historical insight, they conclude by imagining the musings of Marie Antoinette as they might have been in 1792. The queen blames herself for having failed to persuade the king to listen to her advice, and further laments her lack of serious historical study which might have enabled her to advance that advice more authoritatively. Marie Antoinette's poignant wish that she had studied history more systematically seems displaced from her authors who, in their attempt to reclaim historical narrative for women, feel hampered by their lack of access to proper training in research methods. Their ghost story enables them to create their own field of research, constructing the object of the investigation, and subjecting it to the scholarly methods they then evolve.

The subject of this ghostly research is the lost history of women. At the emotional centre of the ghost story is Marie Antoinette's imagined recollection of her beloved Trianon gardens on a particular day in 1789, when she had longed for understanding company, and been briefly conscious of the presence of two sympathetic bystanders. These, the reader perceives, must have been Moberley and Jourdain, who are setting up a two-way current of sympathy with their doomed queen. Her yearnings for her situation to be understood have compelled their time-travel; their sympathetic response justifies their excursion into historical investigation. The peculiarly intense emotionality of the scene derives from the ultimate ineffectuality of their journey; however much they project their sympathy into history, they cannot alter the queen's situation. The ghost story they tell is very similar in structure to Ritchie's opening to *A Book of Sybils*. In each case there is a disruption in time which allows a momentary involuntary encounter between the writer and an apparition from the past, and a potentially powerful exchange of sympathy which is interrupted by circumstance. These late-Victorian and Edwardian ghost stories summarize an urgent wish to import into the past a specifically female sympathy felt to be lacking in most historical reconstruction, and to recover contact with a lost world of strong female feeling, both because that world has been undervalued, and because contact with it might be helpful to women of the present day. Their emotional intensity lies in the pathos attending the difficulties of the attempt. Franco Moretti, in his description of the 'rhetoric of too late' as the major source of tear-jerking effects in writing, points to the crucial moment as the point where the reader's recognition of how things might be changed for the better coincides with the recognition that change is impossible. Tears are produced by the reader's sense of powerlessness and of the irreversibility of time; the ideal situation which produces these tears, Moretti argues, is the death of a character.[17] These women's ghost stories, which promise an apparent expansion of possibilities in the contact between women of different times, only to interrupt that contact, and recognize the irrecoverability of women's history, offer a particularly excruciating version of that pathos.

The recourse to ghost stories marked a particular phase as late-Victorian women writers tentatively explored the possibilities of recording a distinctive women's history. Ritchie's strategies are, however, still traceable in Woolf. In a *Cornhill* review of Ritchie's essay collection, *The Blackstick Papers* (1909), Woolf's exasperation at the scattiness of her aunt's creative methods is tempered by her acknowledgement of the 'magic' in her ability to evoke the past, and in particular to recreate the women of the past:

> The genuine nature of her magic is proved by the truth of much that seems almost too good to be true; she snatches a figure from the past, and shows George Sand; a sort of sphinx in a black silk dress. Her black hair shone dully in the light as she sat motionless; her eyes were fire; it was a dark face, a dark figure in the front of a theatre box.[18]

Woolf's ambivalence about the kind of Victorian female creativity Ritchie represents to her is later elaborated in the figure of Mrs Hilbery in *Night and Day* (1919), composing between bouts of dusting brilliant, incoherent fragments of recollection. Near the end of *Mrs Dalloway* (1925) Mrs Hilbery reappears, offering Clarissa a connection with her hitherto unmentioned dead mother, and commenting on the magic transformation of her garden.[19] Again magic and ghosts are linked with the possibility of recovering a lost and specifically female history. When Woolf looks for women's presence in history, she draws on the specific models Ritchie provided. In her essay on Austen she emphasizes Austen's essential privacy and reserve, and speculates on what she might have written in other circumstances. She also borrows Ritchie's central metaphor of nestbuilding more or less unchanged. 'Humbly and gaily she collected the twigs and straws out of which the nest was to be made and placed them neatly together.'[20] In her most famous construction of tradition, *A Room of One's Own*, Austen appears again as the one fully achieved woman writer in her female disregard for male literary preoccupations, while Eliot fills the Edgeworth role, failing by attempting to insert herself within the male tradition. There is a ghost story in Chapter 1, in which Jane Harrison haunts Newnham gardens, not as the scholar noted for her flamboyant lectures, but as a humble, shabby, tentative figure, offering barely glimpsed evidence of that college's exiguous tradition. The writer who receives most attention is Judith Shakespeare, who carries ghostliness to the point of nonexistence. In the most famous feminist polemic of the earlier twentieth century the ghosts of *A Book of Sybils* still walk.

Notes

1. Anne Thackeray Ritchie, *A Book of Sybils* (London: Smith, Elder, 1883).
2. The essays had appeared separately in the *Cornhill Magazine* in the following order: 'Jane Austen' (August 1871), 'Anna Barbauld' (November 1881), 'Maria Edgeworth' (October–November 1882), 'Amelia Opie' (October 1883).
3. Carolyn Steedman, *Childhood, Culture and Class in Britain: Margaret McMillan, 1860–1931* (London: Virago, 1990), p. 248.
4. Virginia Woolf, 'Lady Ritchie', *Times Literary Supplement* (6 March 1919), p. 23.
5. Robert Louis Stevenson, 'The Faces and the Forms of Yore', in *Complete Poetical Works* (London: Waverley Books, 1924), p. 532.
6. Winifred Gerin, *Anne Thackeray Ritchie* (Oxford: Oxford University Press, 1981), p. 200.
7. Virginia Woolf, *The Complete Shorter Fiction*, ed. Susan Dick (London: Triad Grafton, 1991), p. 78.
8. On the Austen essay's influence see Brian Southam in *Jane Austen: the Critical Heritage*, ed. Brian Southam, 2 vols (London: Routledge and Kegan Paul, 1987), II, 14–26.
9. Gerin, *Anne Thackeray Ritchie*, p. 201.
10. Bonnie G. Smith, 'The Contribution of Women to Modern Historiography in Great Britain, France and the United States, 1750–1940', *American Historical Review*, 89 (1984), 709–32.
11. Margaret Oliphant, *A Beleaguered City* (London: Macmillan, 1930), p. 15.
12. Robert Ackerman, 'Jane Ellen Harrison: the Early Work', *Greek, Roman and Byzantine Studies*, 13 (1972), 209–30; Hugh Lloyd Jones, 'Jane Ellen Harrison 1850–1928', in *Cambridge Women: Twelve Portraits*, eds Edward Shils and Carmen Blacker (Cambridge: Cambridge University Press, 1996), pp. 29–72.

13. Jane Harrison, *Prolegomena to the Study of Greek Religion* (London: Merlin, 1961), p. 273.
14. Jane Harrison, *Reminiscences of a Student's Life* (London: Hogarth Press, 1925), p. 49.
15. Elizabeth Morison and Frances Lamont (pseud.), *An Adventure* (London: Macmillan, 1911). See Terry Castle, *The Apparitional Lesbian: Female Homosexuality and Modern Culture* (New York: University of Columbia Press, 1993), pp. 107–49, for a recent discussion of the Versailles haunting.
16. Joan Evans, 'Introduction', *The Trianon Adventure: a Symposium*, ed. A. O. Gibbons *et al.* (London: Museum Press, 1958), p. 5.
17. Franco Moretti, *Signs Taken for Wonders* (London: Verso, 1983), pp. 159–62.
18. Virginia Woolf, *Essays, 1904–12*, ed. Andrew McNeillie (London: Hogarth Press, 1986), p. 228.
19. Virginia Woolf, *Mrs Dalloway*, eds Elaine Showalter and Stella McNicol (Harmondsworth: Penguin, 1992), pp. 193, 209.
20. Virginia Woolf, *The Common Reader* (London: Hogarth Press, 1962), p. 175.

14
Actresses, Autobiography and the 1890s
Gail Marshall

For much of the nineteenth century, it was the visual impact of the actress's work which attracted most attention. The so-called 'legitimate' and popular theatres alike, cultural spaces often carefully segregated by practitioners and commentators, were united in their practice of relying heavily on the attractive spectacle of the leading lady. Be the play *Romeo and Juliet*, or Madame Vestris' 1830s *Olympian Revels*, the vision of the actress was a crucial element in both the aesthetic and the financial calculations of theatre-managers. As writers such as Bernard Shaw and the drama critic Juliet Pollock noted, such considerations were peculiarly the prerogative of an unsubsidized and therefore necessarily financially speculative theatre, which had to rely upon proven methods of making a profit, and therefore upon the actress to attract audiences. Such was not the case, Pollock noted in the state-supported Théâtre Français.[1] For the actress at the centre of the spectacle, the visual aesthetic curtailed possibilities for self-conscious artistry or for an interpretative negotiation with her play. Like the statue or genre-painting which so strongly informed nineteenth-century staging-practices, the actress was ideally silent in her powers of inspiration and fascination. Even when the actress was intimately involved in the management of the stage and of her own appearance on it, her admission of that activity could only be made retrospectively and often in carefully negotiated terms.

In the 1890s, however, a number of factors combined to alter significantly both the work and the cultural status of the actress, factors which effected a new relationship between actress and audience in particular, and subsequently a new dynamic between the performer and the text. I will seek briefly to isolate these factors, and to suggest the ways in which they enable actresses' participation in an autobiographical aesthetic which has political and professional implications, and which was also the most effective means available of countering her entrapment within the visual. Her engagement with autobiography, I will argue, effectively heralds the demise of the specifically Victorian actress.

The impetus for the new theatrical conditions of the 1890s was given by Janet Achurch's and Charles Charrington's production of *A Doll's House* at the Novelty Theatre in 1889. Ibsen's play was crucial in combining for the first time on an English stage elements which could dispute the dominance of the spectacular aesthetic and liberate the actress from its constraints. The production was financed by pre-paid subscriptions, hence there was no need for the actress to function as a financial lure. Furthermore, the play was an early example of the so-called 'literary drama' which, in its linguistic complexities and sophistication, necessarily involved the actress in a critical reading of the text. She could no longer simply be an appealing image, but had rather to be accepted as an active collaborator in and with the play. Finally, the factor which had most immediate impact upon perceptions of the actress was the nature of Ibsen's heroine herself, and of her story. Nora could not be complacently gazed upon as an attractive young woman, seeming indeed rather to repel the idealizing adjectives and looks usually bestowed upon the young female lead. Indeed, Clement Scott contended that Nora had invented a new 'ideal': she was

> not the pattern woman we have admired in our mothers and our sisters, not the model of unselfishness and charity, but a mass of aggregate conceit and self-sufficiency, who leaves her home and deserts her friendless children because she has *herself* to look after.[2]

In the nature of her struggle and her final door-slamming exit to independence, Nora was participating in both theatrically and politically innovative narratives: she and her creator came to be seen as exemplars of the 'new', 'decadent' movements of the 1890s. In 1894, Hubert Crackanthorpe wrote, 'Decadence, decadence, you are all decadent nowadays. Ibsen, Degas, and the New English Art Club; Zola, Oscar Wilde, and the Second Mrs Tanqueray'.[3] His criticism neatly exposes the coverage of the term, and the extent to which conservative criticism created an homogenized target which misrepresented the multiplicity of the 'New' movement it sought to vilify. However, in the case of the actress it might be argued that such an elision of categories was potentially enabling, newly embedding her as it did within the expediencies and urgencies of the political moment, rather than leaving her stranded within the timelessness of her physical appeal.

One of the most important results of this historicization of the actress was the forging of a link between the 'new' actress and the so-called 'New Women', themselves an eclectic group of writers and political campaigners united in their desire to achieve greater freedoms and equality of opportunity for women. The most immediately apparent connection between the two groups is in their interest in experimenting with new narratives for women's lives, either through acting out those stories, in the plays of Ibsen, Pinero, or actress-playwrights such as Elizabeth Robins, and thus foregrounding the political considerations thus raised; or through writing new narratives of female achievement or newly revelatory stories of despair and failure. Actresses and New Woman writers might also be identified in the public eye as independent, professional women, supporting themselves by means of their own work. However, the possibility of the actress–New Woman link was enabled by, and grounded in, the more fundamental connection of the two groups' very public exploitation of the specifically political import of both the spoken and the written word.

This aspect of the link between actress and New Woman is cemented iconographically in 1894 in *Punch's* satire on the New Woman as 'Donna Quixote', and in the poster for Sydney Grundy's play *The New Woman* (1894), actually an anti-New Woman play. Both images centre upon the figure of a seated woman, one (in *Punch*) an ugly parody of the New Woman as spinster holding a latch-key above her head in what *Punch* obviously sees as a form of redundant triumph; the other a more attractive, younger woman who, during the course of Grundy's play, will be rescued for the cause of matrimony. Both wear pince-nez, and both are framed and defined by the multiplicity of words and books which spill around them, and which are clearly fuelling their errant desires. The books are by authors such as Zola, Ibsen and Tolstoy, and their titles, such as *Man the Destroyer*, clearly reveal apprehensions of the particular threat contained within their covers.

A crucial part of the New Woman writers' strategy was to exploit the political possibilities of the autobiographical genre by telling their own thinly disguised stories in fictional form. Sarah Grand's *The Beth Book* (1898), Ella Hepworth Dixon's *The Story of a Modern Woman* (1894), and Mary Cholmondeley's *Red Pottage* (1899), are good examples of a strategy which played a key role in enabling the development of a collective female, political and aesthetic identity and community. Collective activity operates as both a means and an end in the political arena of the 1890s, and many of the New Woman writers' efforts were concentrated on their effort to draw women together, thus effectively making the act of reading an act of collaborative self-creation and affirmation.

I would like to make a parallel claim for the political function of actresses in the 1890s, and specifically to suggest that their participation, both written and acted, in the art of autobiography, was an act which liberated them from their position of spectacular theatrical attraction, and had the potential to inspire their audiences, like those of the New Woman writers, with new aspirations. As we will see, the actress's access to the autobiographical mode could take a variety of forms. However, no matter how she participates in it, the actress's autobiographical 'act' at this period, which was necessarily both an individual and a potentially collective act, is a mark of her radicalism.

The actress might write her own autobiography; she might be perceived as telling the story of herself, her audience, or both; or her acting might seem to derive from an aesthetic based on an autobiographical principle. A female audience might be able to experience the actress as telling that audience's own story, giving voice to a previously silent narrative in a reciprocal act of narrativization and collective empathy. In *Theatre and Friendship*, the Ibsen-actress, translator and producer Elizabeth Robins recounts how 'One lady of our acquaintance, married and not noticeably unhappy, said laughing, "Hedda is all of us"'.[4] In 1895, Bernard Shaw wrote of Achurch and of the famous Italian tragedienne Eleanora Duse[5] that, 'Every woman who sees Duse play Magda feels that Duse is acting and speaking for her and for all women as they are hardly ever able to speak and act for themselves. The same may be said of Miss Achurch as Nora.' Similarly, William Archer, one of those primarily responsible for the proliferation of the New Drama in England, notes of Duse that 'women should hail her with enthusiasm, for she seems to consummate and ennoble all the nobler possibilities of her sex'; for which ability, a female critic suggests, Duse 'has won the unswerving love of those who know her best, the love and friendship of her own sex'.[6]

Commentators at the end of the century seem to find an autobiographical principle behind the acting of Eleanora Duse in particular. The extent to which, as Helen Zimmern puts it, Duse 'sinks her part in herself rather than herself in her part' became a critical commonplace, and along with her intellect was seen by many commentators as the source of her uniqueness ('Eleanora Duse', p. 990). Her art was 'a part of herself' (Zimmern, 'Eleanora Duse', p. 990), an assessment similar to that made earlier by Archer who writes that Duse's 'most marvellous' creations are 'facets of her natural self' ('Eleanora Duse', p. 305). Indeed Arthur Symons writes that it is even by being 'more and more profoundly herself' that she is most 'profoundly true to the character she is representing'.[7] Duse's art, he continues, is 'wrought outwards from within, not from without inwards' ('Eleanora Duse', p. 202). Duse's acting is thus 'profoundly' autobiographical, telling not of her own experiences, but of her cumulative sympathetic and emotional capacities. Laura Marholm Hansson is eager to claim that this autobiographical source lies in Duse's womanliness, thus making of that art a collective female 'autobiographical' expression:

> There can be no doubt that there is a kind of genius peculiar to women, and it is when a woman is a genius that she is most unlike man, and most womanly; it is then that she creates through the instrumentality of her womanly nature and refined senses. This is the kind of productive faculty which Eleanora Duse possesses to such a high degree.
> 'Eleanora Duse', p. 123[8]

That this impression is 'authored' by Duse rather than by the playwrights she works with is suggested by these critics' perceptions that, without changing the words of the characters she acts, Duse nonetheless changes so radically the conception of her parts that they are scarcely recognizable to the playwright. Of her Marguerite Gautier, critics wrote that 'The picture of manners which Dumas intended disappears in the drama of pure emotion' (Archer, 'Eleanora Duse', p. 302), and that 'she had revealed in [the role] possibilities unknown even to [Dumas]' (Zimmern, 'Eleanora Duse', p. 992). Hansson notes the same of Duse's Nora, 'a being who has no place in the written words, and whom the author never thought of' (*Modern Women*, p. 110). As Zimmern continues, 'Duse collaborates with her authors, and creates their characters anew and afresh' ('Eleanora Duse', pp. 992–3).

Duse is thus thoroughly self-created, emphatically not the result of others' manipulation, neither that of the playwright nor that of the audience, to whom she is rather a 'riddle' (Archer, 'Eleanora Duse', p. 300) than a gratification. The extent of her self-making is perhaps best acknowledged in Archer's and Symons' adoption of the metaphor of sculpting to describe Duse's artistry. Archer writes of her Magda that she is 'a figure, designed and modelled beforehand, proportioned, poised, and polished to the finger-tips with a sculptor's patient assiduity' ('Eleanora Duse', p. 304); while Symons uses the metaphorical likeness between actor and sculptor to assert that through her art Duse even controls 'Nature into the forms of her desire, as the sculptor controls the clay under his fingers' ('Eleanora Duse', p. 196). Symons goes on to elaborate on this function of the actress, and by invoking Rodin, equates the art of his fingers with that of the actress's soul, in which equation of course Duse becomes self-determining, and her body simply her instrument rather than her end:

> The face of Duse is a mask for the tragic passions, a mask which changes from moment to moment, as the soul models the clay of the body after its own changing image. Imagine Rodin at work on a lump of clay. The shapeless thing awakens under his fingers, a vague life creeps into it, hesitating among the forms of life; it is desire, waiting to be born, and it may be born as pity or anguish, love or pride; so fluid is it to the touch, so humbly does it await the accident of choice. The face of Duse is like the clay under the fingers of Rodin.
> 'Eleanora Duse', p. 199

Duse's theatrical body, then, is made not given, and her maker is herself. Symons is thus perhaps the most telling witness of the implications of the autobiographical impulse for the actress. In that impulse there is no space for the moulding hands and gaze of the audience or manager, for their roles have been filled by the actress's self-conscious self-manipulation. Out of that substitution a new critical language, capable of recognizing and of appreciating female 'genius', 'intellect' and 'originality' is born.

Other actresses, who chose to write rather than to act their autobiographies, not only chose their own words, but constructed their whole role through their manipulation of their narratives, thus making a claim for both 'cultural and literary authority'.[9] The autobiographical 'act' of the actress differs significantly from other women's accession to that form. In her professional life, she can already 'command an audience',[10] has already a place in 'the public arena', and is perpetually engaged on-stage in negotiating the acts of 'cultural ventriloquism' which Sidonie Smith suggests are most characteristic of women's struggle with the autobiographical form, and their decision as to the voice in which their story is to be told (*Poetics of Women's Autobiography*, pp. 52, 57). In some respects then, autobiographical writing coincides with the actress's usual practice, but it does so in such a way as to free that practice from its usual constraints within the theatre. The autobiographical subject is both the material of which art is made, and the creator of that art. In self-contemplation, the female autobiographer creates a literary entity out of the self, claiming the right to determine, through authoring, moulding, her own life. The actress's access to the autobiographical impulse, and her newly acknowledged creativity, afford her a similar privilege.

In Helena Faucit Martin's case, this privilege was carefully camouflaged. The actress was best-known for her renderings of Shakespearean women in the 1830s and 1840s, for part of which time she was a member of William Macready's Drury Lane company. First collected in 1885, her autobiographical writings are hidden behind the title *On Some of Shakespeare's Female Characters*, a displacement which perhaps recognizes the implicit claims to authority and self-creation in the act of swapping the stage for the page. In her account of how she became an actress, Faucit Martin invokes a narrative of self-effacement appropriate to the 1880s as she stumbles unselfconsciously into the part of Juliet in the Richmond theatre. She writes of how, as a young girl, on a day when it was too hot to walk by the river, she and her sister stole into the cool, dark theatre whose doors were usually open. Believing themselves to be alone in the place, they acted the balcony scene from *Romeo and Juliet*, with Faucit Martin playing Juliet. Later, they found that they had had a listener:

> When our friends arrived some days later, the lessee told them that, having occasion to go from the dwelling-house to his private box, he had heard voices, listened, and remained during the time of our merry rehearsal. He spoke in such warm terms of the Juliet's voice, its adaptability to the character, her figure, – I was tall for my age, – and so forth, that in the end he prevailed upon my friends to let me make a trial on his stage. [...] Thus did a little frolic prove to be the turning-point of my life.[11]

And thus Faucit Martin's career began. However, her writing herself into that unself-conscious moment, necessarily disrupts the third-party ascription of that unself-consciousness which was a crucial part of the Victorian actress's role. In a highly mediated way she claims autonomy through her self-writing.

The most prolific of the actress–autobiographers of the period was Elizabeth Robins, the first actress to play Rebecca West, Hedda Gabler and Hilda Wangel in English. Her autobiographical writings take a variety of forms and practices. *Both Sides of the Curtain* (1940) is a partial autobiography which submerges the young Robins's attempts to find a place in the London theatre within a history of that theatre in the 1880s. Her own story uneasily infiltrates the more conventional theatre history, just as she herself tried to move onto the London stage, and ends just before she became an Ibsen actress. *Raymond and I* (published posthumously in 1956) is an adventure narrative which takes up Robins's life after her work in the 'new' theatre had finished, and is an account of her journey to Alaska to find and nurse her brother Raymond. The Ibsen years are covered in the essay 'Ibsen and the Actress' (1928), and in *Theatre and Friendship*, an edition of Henry James's letters about the theatre to her and to Florence Bell. Robins's story is told in the commentary which links and contextualizes the letters, and in her strategy of selection which, as Angela V. John notes, demonstrates as great an interest in the story of Robins's collaboration with the recently dead Bell, as in James.[12] The personae in each version are intriguingly varied, from the young ingénue to the pioneer of theatre, from the responsible but adventurous traveller and sister to the writer who ambivalently posed as one in awe of James, 'the Master', whilst, in *Theatre and Friendship*, paying greater tribute to her friend Florence Bell, and taking great delight in evading James's admonitory attentions. In these public texts, Robins self-consciously interrogates the range of roles which were open to her as a woman at the time, and creates and recreates herself in a variety of images.

Crucially, however, in order to achieve this freedom Robins had to leave the stage. By the end of the century, she professes to have come to feel that even Ibsen had let her down. She records of *When We Dead Wake* (1899), that,

> His new play (his last) was a disappointment to some of those who cared most for him. 'When We Dead Awaken' was not, as each of the previous plays had been, a fresh attack; it was full of echoes for the mind and the ear that had much of the older plays by heart. To me the new play was matter almost for tears – no, I would not produce it, even if I were not making ready to go away. The Master hand had weakened, the Master voice was failing.[13]

This is a somewhat startling admission from the actress who had arguably benefited most from the emergence of Ibsen onto the English stage, and who had achieved unprecedented independence as a performer and producer in her work in the New theatre. As she contemplated the end of Ibsen's career as a playwright, she also saw the end of her own life as a performer, believing that she would not be able to go back to the position of 'leading lady' in someone else's company.

Robins's words represent a significant qualification of the freedoms achievable by the actress on-stage, and seem to suggest that even the work of Ibsen was insufficient to overthrow for good the viewing-patterns and prejudices of the Victorian audience. As we have seen, however, for Robins, such a freedom was obtained when she left the stage and took to writing her own narratives. The revelation of the liberating power of language is also at the heart of the Ibsen play which Robins rejected. *When We Dead Wake* shows the former model Irene achieving her liberation from the deadened sculptural form which her erstwhile lover, the sculptor Rubek, had imposed upon her. This is managed through the complementary means of a recourse to a palpably 'symbolic' language which flirts with the anarchy of madness, and through the final vertical movement of the play, which characterizes Ibsen's last works.[14] Irene is liberated finally, of course, into death in the treacherous mountains, but here death acts also as a retrospective entrance into the complexity of a fully realized life which exceeds the death-in-life of the idealized marmoreal form of the model.

Robins's leaving the theatre she had done so much to reform, to take up writing, might seem a similarly complex and paradoxical movement for the actress. It was not, however, one unique to Robins. I will conclude by turning to Ellen Terry, perhaps the most exemplary, and certainly the most popular, of England's Victorian actresses. Even though she did not share Robins's faith in the new theatre movements of the 1890s, and was famously disdainful of Ibsen, she too came to experience a frustration in her function as the chief pictorial attraction of Henry Irving's Lyceum theatre, and to long for greater artistic independence and autonomy. For Terry, too, it is arguably the power of her own autobiographical writing which enables her to enjoy the fulfilment of those desires. A small example of her 'self-writing' beautifully demonstrates this. The actress made her last professional performance with Beerbohm Tree at the Drury Lane Theatre in 1906, when she played Hermione to Tree's Leontes. As Hermione, Terry was far from an exemplary statue. Describing her performance to Graham Robertson, the actress wrote,

> A dreadful thing – I laughed last night!! as the *statue*! and I'm laughing now! Who could help it! With Leontes shouting and Paulina

shouting they just roared so I could not help it! oh, Graham I was rather glad – for I've not been able to laugh at all lately (and *that's downright* wicked) but I'm so afraid I *may do it again* if they will shout!! – You see (as a statue) I don't *look* at 'em – I only hear them – and it's *excruciatingly funny* – I'm *mad* at myself.[15]

Terry's characteristic style and notation struggle to exceed the syntax of her letter, just as her notorious and irrepressible sense of the ridiculous outran the rigours of the beautiful statue's limitations. In some measure, the moment exemplifies the tensions of Terry's career: the conflict between the actress whose beauty was considered her primary theatrical qualification, and the woman whose energy and independence had sustained her in a professional career for fifty years. This incident is particularly interesting, however, in demonstrating the means by which the actress's engagement with writing may elucidate, complicate, elaborate, and finally dispute, her identification with the spectacular statue. Her written account defies the statue's simplicity, its iconic availability and assumptions about its creator, just as Terry's autobiographical writings retrospectively disrupted her spectacular appeal.

Ellen Terry's autobiography, *The Story of My Life*, was not published until 1908, but her first self-writings appeared in the *New Review*, under the disarming title, 'Stray Memories', in 1891.[16] Just as she achieved supremacy on the stage by being perceived as charming, so did Terry adopt a similar practice in these writings, which are a pleasurable, loosely chronological string of anecdotes. However, the crucial factor here is that we are undeniably aware of Terry's decision as an author to be, rather than simply to be found, charming. The act of writing, previously the privilege of her critics, enables her to position her audience, characterizing them in the role of the beloved, to be charmed and wooed by her text, and subsequently, to have their reactions influenced, if not determined, by the decisions that she as a writer is making. Terry's rhetoric of charm functions skilfully to camouflage the substance of her text; it acts as a substitute for the romantic interest which is entirely lacking in her writing, and detracts attention from, or at least mollifies apprehensions of, the almost entirely professional narrative she writes.

Her early memoirs analyse Terry's career in such a way as scarcely to mention Irving, or the other men with whom she was associated. She omits any mention at all of G. F. Watts and Edward Godwin, her relationships with whom were well-known, and the two men are engulfed anonymously in the textual gaps of her absences from the stage. The articles are shrewdly constructed accounts of the actress's training, leavened with the kind of personal reminiscence associated with theatrical memoirs. But even the latter contain appreciations of particular skills, such as Mr Kean's voice projection ('Stray Memories', p. 337), or lead to reflection on theatrical practices. Terry is effectively outlining how she had moulded her own career, acknowledging as a mentor only her early teacher, Mrs Kean ('Stray Memories', p. 333). These are professional rather than personal memoirs, and in their autobiographical matter, as well as their form, publicly constitute Terry's claim to professional autonomy.

For Terry, writing autobiographically is not simply a matter of describing 'how the author acquired or came to do without a room of her own; how she came to command an audience rhetorically, ideologically and socio-economically' (Broughton, 'Women's Autobiography', p. 79). Rather the act constitutes her claim to order, and have ordered, both her narrative and the life she describes. Such a strategy, later more fully played out in *The Story of My Life*, is not the act of a submissive or reflexive woman, whose identity is constituted by another. For the actress to enter into the world of authorship is to enter upon new possibilities of self-constitution, and specifically to envisage new temporal possibilities for the persistence of the evidence of that self. Reliance upon the body and its impact is surpassed in the act of writing. This disturbing effect is aided by the articles' lack of illustrations, which further isolates and protects her words from the connotations of Terry's body.

In 1891, the same year as 'Stray Memories' were published, Terry appeared with her son Gordon Craig in an adaptation of Charles Reade's curtain-raiser, *Nance Oldfield*, based on the life of the famous eighteenth-century actress. When it was originally offered to the Lyceum, Irving turned the play down, and it was bought instead by Terry. The actress was a key participant in this theatrical event which gave a new lease of life to one of her professional ancestors, and in the course of which Terry both asserted her awareness of her own place within a historical community of actresses and tasted the power of the playwright and producer in collaborating with Bram Stoker in the adaptation of the play. *Nance Oldfield* opened on 12 May and became a stock part of Terry's repertoire. Far from being a simple vehicle for displaying Ellen Terry, the play may be read both as an act of communication with the long-dead actress, and as Terry's comment on her role and place as a contemporary actress, told through her participation in a homage to the famous eighteenth-century actress Ann Oldfield. The play signals a degree of self-consciousness on Terry's part which militates against the easy assumption that she is created simply through the desires of her spectators. Nance Oldfield's reflections on acting necessitate the recognition of Terry's own adoption of a professional role, as, for instance, when Nance reveals what she feels when acting on-stage:

> I ride on the whirl-wind of a poet's words; I wave my mimic sceptre with more power then the kings of Europe – they govern millions of bodies, but I sway a thousand hearts – and the next day I sink to woman![17]

Further, in most self-conscious mode, Terry parodies her own roles in the plays of 'Willy Shakespeare', with brief renditions of Lady Macbeth and Juliet. Her acting edition of the play is full of such amendments to her part. In this play, Terry fundamentally challenges the parameters of her role as the Lyceum's leading-lady, for in many of her activities, and especially as adaptor and owner of the play, she was usurping Irving's more usual responsibilities. Terry uses the role to tell her own professional story, to narrate her own understanding of her part in the theatre, and thus to write another form of autobiography.

Terry's autobiography proper has had an often controversial history, with both her children, Edy and Gordon Craig, showing themselves concerned in some measure to re-write the character Terry creates of herself. In *Ellen Terry and her Secret Self*, Gordon Craig presents Terry as fundamentally fractured, living a dual existence as both 'that little Nelly who was my mother – her secret self – [... and] that redoubtable adversary, her other self, Ellen Terry', the successful, professional actress.[18] As part of their own negotiation with both Gordon and Ellen herself, Edy Craig along with her female partner Christopher St John, republished the autobiography under the title *Ellen Terry's Memoirs* in 1933, a version which professed to have made few alterations to the original text, but which had in fact edited it heavily, and added weighty footnotes and highly partisan chapters detailing the final years of Terry's life, which irredeemably alter the nature and generosity of the actress's reminiscences.[19]

As Craig and St John had noted in their commentary, they wanted to construct a written record which would allow Terry's fame to endure beyond 'the memory of a diminishing band of elderly people' (*Ellen Terry's Memoirs*, p. 348). As they added later, 'Iliads are more enduring than Parthenons' (*Ellen Terry's Memoirs*, p. 354). However, from the brief evidence given here of the siblings' struggle with their mother's memory, and as the number of works written about Terry demonstrate, it would seem that the most important and liberating aspect of self-writing for the actress is not the endurance promised in St John's and Craig's classical allusions, but rather her entry into the realm of interpretation and consequent mutability and variety. The too-solicitous St John misses the point of Terry's writing, which, rather than substituting one lapidary form for another, instead inserts Terry into the variety and complexity of the textual witness which is profoundly at odds with the dimensions of a commemorative monument.

Writing of Terry's *The Story of My Life* in her 1941 essay on the actress, Virginia Woolf suggests that with her pen Terry has 'painted a self-portrait', but that this is 'not an Academy portrait, glazed, framed, complete'.[20] She goes on in terms which appropriately both dismember the actress's remembered and much-adored body, and reconstitute that body as a self-constructed literary artefact:

> It is rather a bundle of loose leaves upon each of which she has dashed off a sketch for a portrait – here a nose, here an arm, here a foot, and there a mere scribble in the margin. The sketches done in different moods, from different angles, sometimes contradict each other. The nose cannot belong to the eyes; the arm is out of all proportion to the foot. It is difficult to assemble them.
>
> Woolf, 'Ellen Terry', p. 68

Difficult, that is, to make them cohere in a stable form. Writing herself has enabled Terry to elude her audiences by forcing them to engage with her in the medium of her own language which was usually so disablingly denied to the Victorian actress, and which has enabled this actress to survive beyond the moment at which she was apparently brought into being by the attentions of her audience.

Notes

A version of this essay first appeared in Chapter 5 of my book, *Actresses on the Victorian Stage: Feminine Performance and the Galatea Myth* (Cambridge: Cambridge University Press, 1998) ('Living Statues and the Literary Drama'). I am grateful to Cambridge University Press for permission to reproduce extracts here.

1. Juliet Pollock, 'The Comédie Française', *Contemporary Review*, 18 (1871), 43–55.
2. Clement Scott, 'A Doll's House', *Theatre*, 14 (1889), 19–22; quoted in *Ibsen: the Critical Heritage*, ed. Michael Egan (London: Routledge and Kegan Paul, 1972), p. 114.

Actresses, Autobiography and the 1890s 221

3. Hubert Crackanthorpe, 'Reticence in Literature', *Yellow Book*, 2 (1894), 259–69 (p. 266).
4. Elizabeth Robins, *Theatre and Friendship: Some Henry James Letters* (London: Cape, 1932), p. 18.
5. Duse was, along with Ellen Terry and Sarah Bernhardt, one of the three leading actresses of the international stage in the 1890s. Far from falling into the pattern of 'rival queens' constructed for them by their critics, the three women seem to have enjoyed a relationship based on mutual respect and even affection.
6. Bernard Shaw, 'Toujours Daly' (13 July 1895), in *Our Theatre in the Nineties*, 3 vols (London: Constable, 1932), I, 177–84 (p. 183); William Archer, 'Eleanora Duse', *Fortnightly Review*, 58 (1895), 299–307 (p. 301); Helen Zimmern, 'Eleanora Duse', *Fortnightly Review*, 67 (1900), 980–93 (p. 993).
7. Arthur Symons, 'Eleanora Duse', *Contemporary Review*, 78 (1900), 196–202 (p. 201).
8. An alternative derivation for Duse's art, and one which also explicitly exceeds the parameters of the visual aesthetic, is suggested by Vernon Blackburn, who suggests that, 'If you subtract the imaginative creativeness from each character which she impersonates, you are left with a quality of pure intellect, in each instance. This is the explanation of her everlasting variety and of the singleness of her simplicity' – Vernon Blackburn, 'Eleanora Duse', *New Review*, 13 (1895), 39–44 (p. 41).
9. Sidonie Smith, *A Poetics of Women's Autobiography: Marginality and the Fictions of Self-Representation* (Bloomington: Indiana University Press, 1987), p. 50.
10. T. L. Broughton, 'Women's Autobiography: The Self at Stake?', *Prose Studies*, 14 (1991), 76–94 (p. 79).
11. Helena Faucit, Lady Martin, *On Some of Shakespeare's Female Characters* (Edinburgh: Blackwood, 1885), p. 90.
12. See Angela V. John, *Elizabeth Robins: Staging a Life* (London: Routledge, 1995), p. 96.
13. Elizabeth Robins, *Raymond and I* (London: Hogarth, 1956), p. 47.
14. Inga-Stina Ewbank, 'The Last Plays', in *The Cambridge Companion to Ibsen*, ed. James McFarlane (Cambridge: Cambridge University Press, 1994), pp. 126–54.
15. Letter of 6 September 1906; quoted in Nina Auerbach, *Ellen Terry: Player in Her Time* (London: Dent, 1987), p. 283.
16. Ellen Terry, 'Stray Memories', *New Review*, 4 (1891), 332–41; 444–49; 499–507 (p. 333).
17. Ellen Terry's acting copy of *Nance Oldfield*, p. 6. The edition is kept at the National Trust's Ellen Terry Memorial Museum, Smallythe Place, Kent.
18. Edward Gordon Craig, 'Preface' to *Ellen Terry and Her Secret Self* (London: Sampson Low, 1932), pp. vii–viii.
19. *Ellen Terry's Memoirs*, with Preface, Notes and Additional Biographical Chapters by Edith Craig and Christopher St John (London: Gollancz, 1933).
20. Virginia Woolf, 'Ellen Terry', in *Collected Essays*, 4 vols (London: Hogarth, 1967), IV, 67–72 (p. 68).

15
Modernity and Progress in Economics and Aesthetics

Regenia Gagnier

A recent book on Oscar Wilde's critical writings develops the thesis that the *fin-de-siècle* Decadence *is* modernity in that it seeks to contain within itself all the moods and forms of the past.[1] The Decadent critic who knows the many moods and modes of the past is a 'personality' whose power comes from the masks through which he fashions and refashions himself (for reasons that will become clear below, in the nineteenth century it was generally 'he'). This reference to the past, or what the author in question, Lawrence Danson, calls the Decadence's 'backwardness', challenges Victorian notions of progress: hence its 'modernity'. Similarly the Decadent idea of personality as multiplicity and surface rather than soul within challenges the earnest Victorian idea of the singular and autonomous individual: hence *its* 'modernity'.

This view of modernity as summing up in itself all prior modes of thought and life is not a new one. In his essay on Leonardo Da Vinci, Walter Pater called it *'the* modern idea'. 'All the thoughts and experience of the world have etched and moulded' in La Gioconda's face,

> The animalism of Greece, the lust of Rome, the reverie of the middle age with its spiritual ambition and imaginative loves, the return of the Pagan world, the sins of the Borgias [....] [L]ike the vampire, she has been dead many times and learned the secrets of the grave; and has been a diver in deep seas, and keeps their fallen day about her; and trafficked for strange webs with Eastern merchants; and, as Leda, was the mother of Helen of Troy, and as Saint Anne, the mother of Mary.[2]

Pater concludes the famous passage with the modernity of the Darwinian idea of evolution, 'modern thought has conceived the idea of humanity as wrought upon by, and summing up in itself, all modes of thought and life'.

Whereas Danson, in his book on Wilde's criticism, would stress (after the manner of Paul de Man)[3] modernity's obsession with the past in its 'summing up' of all previous modes and moods, I have stressed the 'summing up' itself, or modernity's capacity to consume randomly all experience. Pater's stunning ahistoricism in the juxtapositions and appropriations of *The Renaissance* (1873; 1893) exemplifies the modern (Decadent) critic's capacity to consume the treasures of the past as contributing to his own 'unique' 'personality', revealed by his distinctive tastes, or even by his psychological dispositions. For of course the psychological summing up – the layering of prior experience in the psyche – is also figured in La Gioconda's image and being formulated as modern at the *fin de siècle*. The paradox, of course, is that the critic or 'personality' who objectifies and consumes the exotica of the world in the service of establishing his own taste or discrimination is doomed by his own publicity: the individual in the age of 'personality' becomes a stereotype, a representative of a class. Thus Pater becomes the 'aesthete' and his tastes 'aestheticism', Wilde becomes the homosexual or an 'Oscar', Beardsley becomes 'Beardsley', or the Mona Lisa becomes the standard representative of 'masterpiece' in the age of mechanical reproduction.

The key elements of modernity, then, are consumption, not only of time as in the past, but also of space, as in Decadent exoticism and its correlative fetishism or, as Anne McClintock calls it, 'commodity racism';[4] and, second, a kind of individual or self produced by this ability to consume time and space. Recall that what is central to Pater's description of the quintessentially modern Mona Lisa is *desire*: her image 'is expressive of what in the ways of a thousand years men had come to desire' (*Selected Writings*, p.46). Like the vampire, she has consumed the world, 'trafficked for strange webs with Eastern merchants', and we have come to desire her for her 'beauty wrought out from within upon the flesh, the deposit, little cell by cell, of strange thoughts and fantastic reveries and exquisite passions' (Pater, *Selected Writings*, p. 46). 'Lady Lisa', says Pater, 'might stand as [...] the symbol of the modern idea' *(Selected Writings*, p. 47), for consumption is the essence of modernity. This is a truism of Western economics and political economy, most uncompromisingly formulated in W. W. Rostow's 'stages of economic growth', whose final stage, the end of history, culminated in American-style 'high mass consumption'.[5]

This essay will consider the cultural appurtenances of the economic theory, in the first case describing the shift from production to consumption models in economics and aesthetics in the second half of the nineteenth century, and then focusing on the 'civilized' modernity of the man of taste and his implication in the European or white discourse of comparative civilizations.

The aesthetics of consumption in the last quarter of the nineteenth century were concurrent with the economics of consumption. The economics culminated in marginal utility theory and a view of economic man as consumer that displaced classical political economy's notions of economic man as producer (Smith, Ricardo, Mill and Marx) or reproducer (Malthus). This shift from economic man and woman as producer and reproducer to economic man and woman as consumer gave rise to competing politics of labour and desire. The story, briefly, goes like this.[6] The essence of political economy as it was practised from the later eighteenth to the mid-nineteenth centuries was social relations, hence the term 'political'. To a culturalist, one of the most important insights of classical political economy, say from Smith to Mill, was that the division of labour was the source of differences between people. People may or may not identify with a social or economic class; in Britain in the nineteenth century they often did, in the US today they typically do not. But most people's subjective and objective identities are centrally related to whether they make nails, automobiles, books, contracts, breakfast, hotel beds, music, speeches or babies. The fact that the division of labour also reflects major social divisions of race, gender and ethnicity, and internationally reflects relations of domination and subordination between nations, is also crucial in establishing individual and collective identities.

Second, the political economists were concerned about the negative consequences of the division of labour. Mixing mechanistic and organic metaphors, Adam Smith proposed government mechanisms to ameliorate British workers' deterioration in what he called the social, intellectual, and martial virtues.[7] Mixing market and virtue ideology, John Stuart Mill feared that competitive individualism would drive out sympathy and altruism.[8] And Marx and Engels, who criticized political economy while adopting some of its fundamental categories, put alienation and atomism, respectively, at the centre of working and bourgeois life.[9] Despite their penchant for economic laws and the abstraction of a self-interested maximizer of material advantage called Economic Man, the political economists also believed that economic systems made kinds of people and that the division of labour, as another critic, John Ruskin, said, also divided people from one another.[10]

Third, the political economists, being typically *progressive* rather than (like Rostow) *developmental* – I shall say more about this distinction below – did not believe that markets were the end of history. Markets were taken as one stage of growth, but economic growth was no more an end in itself than Beauty was to their contemporaries in aesthetics. Smith thought that free trade, if it ever happened (which he thought unlikely), would lead to world peace (the so-called Doux–commerce thesis). Mill thought that once production reached a certain level, society's primary concern should be more equal distribution and indeed thought that the appropriate level of production had already been reached in 1871, which inclined him toward socialism late in his life. Political economy entailed a theory of social relations in a world in which scarcity was perceived to be a relationship of productive forces to Nature, and markets were appropriate to but one stage of the development of those productive forces. Once society had *developed* its productive forces, humanity could *progress* ethically and politically.

Around 1871, economic theory began to shift its focus from the social relations of population growth, landlords, entrepreneurs, workers and international trade to the individual's subjective demands for goods. The labour theory of value, which had seen the human body and human labour as the ultimate determinants of price, was abandoned in favour of consumer demand. Value no longer inhered in goods themselves – whether the goods were grain or human labour – but in others' demand for the goods. Political economy's theory of the productive relations between land, labour and capital thus gave way to the statistical analysis of price lists or consumption patterns. One of the corollaries of marginal utility theory, as it came to be called, was that consumer choice ceased to be a moral category: it did not matter whether the good desired was good or bad, just that the consumer was willing to pay for it. Value ceased to be evaluative across persons: it became individual, subjective, or psychological.

The psychological bias transformed the modern concept of scarcity. For Malthus and the political economists, scarcity was a relation of productive forces to the earth, as in population growth and diminishing returns from agriculture. Under marginal utility theory, scarcity was relocated in the mind itself, as a consequence of the insatiability of human desires. Stanley Jevons, Carl Menger, and the other early theorists of consumption saw that as the basic needs of subsistence were satisfied, humankind's desire for variety in shelter, food, dress and leisure grew limitlessly, and thus the idea of needs – which were finite and the focus of political economy – was displaced by the idea of tastes, which were theoretically infinite.[11]

All this amounted to a noticeable shift in the concept of economic man. Under political economy, economic man was a productive pursuer of gain; for Jevons and Menger, on the other hand, economic man was a consumer, ranking his preferences and choosing among scarce resources. Significantly, *modern* man would henceforth be known by the insatiability of his desires, and the indolent races of savages – whether Irish, African, or native American (key examples of the period) – needed only to be inspired by envy to desire his desires, imitate his wants, to be on the road to his progress and his *civilization*. (I shall say more about this term below.) His nature, insatiability, was henceforth human nature itself. His mode, consumer society, was no longer one stage of human progress, but its culmination and end. This was the burden of the 'comparative civilizations' histories of the period. Ideas of *progress*, which implied moral and political progress as well as economic growth, began to be displaced by ideas of *development*, which implied only an inevitable trajectory toward high mass consumption. Here I shall focus on the man of taste in relation to the savage, which represented the relation of Britain to external Others. Yet equally important in Britain was the man of taste in relation to the Barbarian, generally an internal figure, as in Matthew Arnold's distinctions in *Culture and Anarchy* (1869) between 'Barbarians, Philistines, and Populace'. Unlike the savage, who had not achieved the higher orders of civilization, the Barbarian had wilfully resisted civilization. Generally representing a degenerate aristocracy, as in Arnold, the Barbarian did not exercize the self-restraint of the middle-class ('Philistine') man of taste. The man of taste distinguished himself equally from the simple savage and the degenerate barbarian.

I have shown in my recent research how a shift from a production ethos to a consumption ethos also occurred in aesthetics. That is, an aesthetics that was concerned with ethics or social relations, and production or creativity, by the end of the nineteenth century had begun to shift its focus from the work of art and the artist to the consumer of art, the so-called man of taste, or the critic.[12] The contours of the shift are evident in popular aesthetics, or the domain of fiction. Nineteenth-century Britain was the great era of the novel as a form, and the fiction of Jane Austen, Mary Shelley, Elizabeth Gaskell, George Eliot, the Brontës, Charles Dickens, William Thackeray, Anthony Trollope, and the rest was the fiction of social relations. Political economy saw the world in terms of the three perceived classes of the time (landowners, workers and entrepreneurs) and their commodified objects of exchange (land, labour and capital), resulting in rent, wages and profits in domestic and colonial markets. Nineteenth-century fiction correspondingly represented the relations between landed aristocrats, entrepreneurs and wage earners. Whether the aristocrats were represented as stable pillars of the community (as in Trollope) or lazy and decadent (as in Bulwer-Lytton); whether the entrepreneurs were energetic (as in Gaskell) or cruel (as in Dickens); whether the wage-earners were docile and dependent (as in Eliot) or angry and seditious (as in Disraeli), depended on the political perspective of each novelist. But there is no doubt that the great novelists saw the world in terms of social groups in contact and often in conflict. Their view of socio-economic relations extended considerably beyond that of the political economists, who refused to acknowledge the arena of unpaid work, like much housework and care of dependents, or widespread but illegitimate work, like prostitution. The novelists did not have a limited view of the economy. To the contrary, even the fiction of greatest psychological depth, like Brontë's *Villette* (1853), finds economic relations constitutive of the psyche.

In the area of aesthetic *theory*, political economy's emphases on production and social relations may best be illustrated by the practices of those John Ruskin called the Political Economists of Art.[13] The political economists of art were centrally concerned with producers or creators of wealth, but they defined wealth differently from the political economists. Rather than economic growth alone, they sought to provide the conditions for producers whose work would be emotionally, intellectually and sensuously fulfilling, and whose societies would be judged by their success in cultivating creators and creativity, very broadly conceived. William Morris, a poet, novelist and master craftsman, repeatedly denied the autonomy of art, claiming first that 'art should be a help and solace to the daily life of all men' and extended art's arena 'beyond those matters which are consciously works of art [...] to all the externals of our life'.[14] He demanded on behalf of art good health, to the extent that society could provide it; liberating sensuous experience (that is, freedom from shame, to the extent that society could provide it); education to the extent of an individual's capacity in both knowledge and skill of hand; and, finally, the right to reject what he and Ruskin called 'destructive' forms of work, specifically war and drudgery, or work that did not employ the skill or imagination of the worker.

Central to the political economy of art was a labour theory of value, or a theory of the creative production *and reproduction* of warm human bodies. By the *fin de siècle* such productivist theories were in competition with theories of desire and consumption. Where the political economists of art had emphasized the creativity of the artist and builder, and high Victorian fiction had emphasized women's role in the reproduction of the domus, the *fin de siècle* emphasized the consumer of the work and the incessant search for pleasure and sensation. In the France that so influenced the British 1890s, Baudelaire had consciously reacted against an ethos of productivity, going so far as to reject masculine virility itself. The more a man cultivates taste, he wrote approvingly, 'the less often he gets an erection [...]. Only the brute gets really good erections'.[15] The hero of J.-K. Huysmans's novel *A Rebours* (1884) gave himself, we read, a funeral banquet in memory of his own virility, 'lately deceased'.[16] George Moore's hero Mike Fletcher in the novel of that name treated women like cigarettes, consuming and disposing of them in an insatiable search for stimulation:

> More than ever did he seek women, urged by a nervous erithism which he could not explain or control. Married women and young girls came to him from drawing-rooms, actresses from theatres, shopgirls from the streets, and though seemingly all were as unimportant and accidental as the cigarettes he smoked, each was a drop in the ocean of the immense ennui accumulating in his soul.[17]

Oscar Wilde's description of a cigarette also described the perfect commodity: cigarettes, Wilde said, were the perfect type of the perfect pleasure, because they left one unsatisfied.[18] The *fin de siècle*'s basic stances toward the economy – boredom with production but love of comfort, insatiable desire for new sensation, and fear of falling behind the competition – culminated in Max Beerbohm's publication of his *Complete Works* at the age of twenty-four. 'I shall write no more', he wrote in the Preface of 1895, 'Already I feel myself to be a trifle outmoded. I belong to the Beardsley period. Younger men, with months of activity before them [...] have pressed forward since then. *Cedo junioribus.*' Beerbohm satirized the duality of aestheticizing/commodifying one's life in *Zuleika Dobson* (1911), in the double images of dandy and female superstar. Real women, like Mrs (M. E.) Haweis in her *Beautiful Houses* (1881), on the other hand, were packaging the world in moments of taste and connoisseurship and commodifying them for suburban effects. Examples in aesthetic theory of what the late

Victorians called the 'hedonic calculus', or the aesthetics of consumption, pleasure and taste, pervade the *fin de siècle* and I've discussed many of them elsewhere. I've also written at length about how a focus on women writers and artisans complicates the Decadents' aesthetic of consumption and shows the tensions between production and consumption models within different social groups.[19]

In the relationship of this notion of modernity as the individual's capacity to consume to notions of progress, paramount was the idea that all progress was progress toward individualism. The Victorian search for a law of progress was most pronounced in Herbert Spencer's *Social Statics* (1851) and *Progress: Its Law and Cause* (1857). Our organs, faculties, powers and capacities grow by use and diminish from disuse, and we may infer, according to Spencer, that they will continue to do so. 'Thus the ultimate development of the ideal man is logically certain'; humanity must in the end become completely adapted to its conditions:

> Progress, therefore, is not an accident, but a necessity. Instead of civilization being artificial, it is a part of nature; all of a piece with the development of the embryo or the unfolding of a flower [....] As surely as the tree becomes bulky when it stands alone, and slender if one of a group; as surely as the same creature assumes the different forms of cart-horse and race-horse, according as its habits demand strength or speed; as surely as a blacksmith's arm grows large, and the skin of a labourer's hand thick; as surely as the eye tends to become long-sighted in the sailor, and short-sighted in the student; as surely as the blind attain a more delicate sense of touch; as surely as a clerk acquires rapidity in writing and calculation; as surely as the musician learns to detect an error of a semi-tone amidst what seems to others a very babel of sounds; as surely as a passion grows by indulgence and diminishes when restrained; as surely as a disregarded conscience becomes inert, and one that is obeyed active; as surely as there is any efficacy in educational culture, or any meaning in such terms as habit, custom, practice; so surely must the human faculties be moulded into complete fitness for the social state; so surely must the things we call evil and immorality disappear; so surely must man become perfect.[20]

This picture of perfect adaptation in Spencer becomes increasingly organic, so that the law of organic progress, consisting in the change from the homogeneous to the heterogeneous, individuated, or unique, is revealed as the law of all progress:

> The investigations of Wolff, Goethe, and von Baer, have established the truth that the series of changes gone through during the development of a seed into a tree, or an ovum into an animal, constitute an advance from homogeneity of structure to heterogeneity of structure. In its primary stage, every germ consists of a substance that is uniform throughout, both in texture and chemical composition. The first step is the appearance of a difference between two parts of this substance; or, as the phenomenon is called in physiological language, a differentiation. Each of these differentiated divisions presently begins itself to exhibit some contrast of parts: and by and by these secondary differentiations become as definite as the original one.
>
> Spencer, Progress, p.436

Spencer conjectures that the scope of the process is literally universal:

> If the nebular hypothesis be true, the genesis of the solar system supplies one illustration of this law [....] Whether it be in the development of the earth, in the development of life upon its surface, in the development of society, of government, of manufactures, of commerce, of language, literature, science, art, this same evolution of the simple into the complex, through successive differentiations, holds throughout.
>
> Progress, p. 436

His examples of increasing complexity include the division of labour, the global market, languages, human physiology, and transnational types – the European is more heterogeneous or individual than the Australian, the Anglo-American the most heterogeneous or individual, and therefore the most advanced, of all. (We shall address the non-national bias of such claims below: civilization is European or white, not national.) Spencer's explanation of this universal transformation of the homogeneous into the heterogeneous is the history of multiple effects from singular causes: 'Every active force produces more than one change – every cause produces more than one effect [....] From the law [...] it is an inevitable corollary that during the past there has been an ever-growing complication of things' (*Progress*, p. 442).

From the early political economists' view of a homogeneous human nature, in which variation is caused by the division of labour and accidents of location, we have moved to, with Spencer, heterogeneous populations, the most élite of which were increasingly differentiated and complex. From the historical causes of variation or heterogeneity in specific populations, we have moved to the laws of variation or heterogeneity. Under the influence of anthropology, Spencer biologized the division of labour, making differences between people evolutionary, or organically purposive.[21] The logic of his system with respect to what he called the 'higher races' was toward increasing individualism, voluntary co-operation, and mutual aid in a division of labour and markets.

With respect to the lowest races, as he called them in *Descriptive Sociology* (1874), the logic of his system converged with evangelical conceptions, in which, for example, savages and barbarians acted upon impulse for immediate gratification, whereas civilized man's instincts were modified by reason. Thus, unlike the savage or barbarian, economic man's aversion to labour was offset by his desire for wealth, or, in his sexual economy, his instinct for immediate gratification was offset by the sublimation of his sexual appetite. This cultural evolutionary variation of biological determinism, in which what was biologically hereditary could itself be the result of cultural processes, was compatible with political economy's notions of restraint, abstinence or saving (Marx famously called attention to political economy's paradoxical status at that time as both the science of wealth and 'the science of renunciation').[22]

Consider what Spencer calls 'civilized' man. The man of taste was quintessentially the civilized man, and the civilized man had a history in the discourse of civilizations that put him in dialectical relation with savage or barbarous Others, who in addition to well-defined deprivations of culture also lacked in taste or discrimination.[23] In the nineteenth century, the idea of progress was inextricably linked with the idea of civilization. This has been summarized by the geographer Peter Taylor:

> Social science evolved at the same time that Europe finally confirmed its dominium over the rest of the world. This gave rise to the obvious question: why was this small part of the world able to defeat all rivals and impose its will on the Americas, Africa, and Asia? [...] [A]nswers to [this question] were found not at the scale of the state but in terms of civilizations. It was Europe as 'Western' or 'modern' civilization that had outgunned and outproduced all-comers, not Britain or France or Germany despite the sizes of their individual empires. The domination was a collective European enterprise, not a particular state enterprise. This concern with how Europe did it coincided with the major intellectual transition associated with Charles Darwin [....] Evolutionary theory was popular in

early social science because it provided scientific legitimation to the assumption of progress that culminated in the self-evident superiority of contemporary European society. Hence sociologists and anthropologists could produce their famous stage theories of social evolution ending in industrial civilization. Historians could identify paths of progress to modern civilization and the many cul-de-sacs of civilizations not blessed with the secret of progress [....] Geographers joined in with their theories of geographical determinism to show that Europe and European-settler regions had the necessary climate to stimulate ever higher levels of civilizations, leaving other less fortunate civilizations 'immobile'.[24]

Taylor calls these studies 'comparative civilization' studies. They included the global evolutionary comparisons of armchair anthropologists and historians working with newly generated statistical data, though they had roots in the universal histories of the Enlightenment.[25] Their function was to interpret the past and indicate improvement, to evaluate difference and propose that peoples could be ranked along a common course (the evolving Family of Man), and to supply an optimistic vision of human destiny and a faith in the ameliorative power of change. In them, the idea of European or white 'civilization' implies civil society, private property, the social order and, most importantly for us, refinement or taste at the individual level. The 'civilized man' or Spencer's individual is their apex. Consider the world-historical formulations:

> Shall all nations someday approach the state of civilization attained by the most enlightened, the freest, the most emancipated from prejudices of present-day peoples, such as the French, for example, and the Anglo-Americans? Shall not the vast interval which separates these peoples from the bondage of nations subservient to kings, from the barbarism of the African tribes and the ignorance of savages, gradually disappear? Do there exist on the globe countries whose inhabitants nature has condemned never to enjoy liberty, never to exercise their reason?
> Condorcet, *An Historical Picture of the Progress of the Human Mind* (1795)[26]

> The history of the world travels from east to west, for Europe is absolutely the end of history, Asia the beginning [.... F]or although the earth forms a sphere, history performs no circle around it, but

has on the contrary, a determinate east, viz., Asia. Here rises the outward physical sun, and in the west it sinks down: here [in the West...] rises the sun of self-consciousness, which diffuses a nobler brilliance. The history of the world is the discipline of the uncontrolled natural will, bringing it into obedience to a universal principle and conferring subjective freedom. The east knew and to the present day knows only that *one* is free; the Greek and Roman world, that *some* are free; the German world knows that *all* are free.

Hegel, *Philosophy of History* (1837)[27]

[It is only] the selectest part, the vanguard of the human race, that we have to study; the greater part of the white race, or the European nations, – even restricting ourselves, at least in regard to modern times, to the nations of Western Europe [....] In short, we are here concerned only with social phenomena which have influenced, more or less, the gradual disclosure of the connected phases that have brought up mankind to its existing state. When we have learned what to look for from the élite of humanity, we shall know how the superior portion should intervene for the advantage of the inferior.

Comte, *Postive Philosophy* (1830–42)[28]

Comte ultimately believed in political laissez-faire:

It is not our business to suppose that each race or nation must imitate in all particulars the mode of progression of those who have gone before. Except for the maintenance of general peace, or the natural extension of industrial relations [a Big Except], Western Europe must avoid any large political intervention in the East; and there is as much to be done at home as can occupy all the faculties of the most advanced portion of the human race.

Comte, *Postive Philosophy*, p. 396

Such assumptions as these pervade economic thought of the nineteenth century and are inextricable from notions of economic man as producer and, later, as consumer, culminating in the European man of taste.

This idea of civilized man – as in 'that's damned civilized of you' (notice how it is parallel with, in the US, 'that's damned white of you') – everywhere pervades late Victorian aesthetics of taste. And here the common speech of 'civilization' is even more significant than the grand theories just cited.[29] From Disraeli's *Tancred* (1847; the secret of which is a white queen of Arabia), to Haggard's *She* (1887; the secret of which is a white queen of Africa), the empire of literature was tantamount to the literature of empire in its obsession with comparative civilizations and appropriation of popular evolutionary models. By the end of the nineteenth century, however, the narratives of progress were less likely to terminate in Reason or Freedom in their Hegelian or Kantian senses than in the sublime and *irrational* images of devolution or fear of engulfment in the late Victorian Gothic or economic rationality and the individual's freedom to maximize self-interest. And we should note that 'interest' itself is a psychological term that in the nineteenth century was distinguished from the social good, the social body, or even sociology. As psychological models were replacing the sociological by the *fin de siècle* in both aesthetics and economics, deep, disturbed unreason, on the one hand, or economic rationality and instrumental reason, on the other, were replacing Reason as the mind's divine capacity to improve its own condition and that of others. I have written about this in relation to the psychological aesthetics and psychological fiction of the *fin de siècle*, which in both cases calculated pleasure and pain in the pursuit of happiness as self-interest (which was very different from Bentham's utilitarian pursuit of happiness). The economics were formalized in Richard Jenning's physiological psychology, which underlay Jevons's marginal utility theory, and Francis Ysidro Edgeworth's *Mathematical Psychics* (1881). In the latter work Edgeworth recommended that, since the capacity for pleasure had evolved such that men and Europeans had a greater capacity for pleasure than women and non-Europeans, 'the comfort of a limited number' could mathematically outweigh 'the numbers with limited comfort' when it came to issues of distribution. One may well see here how the calculations of taste implied a global and gender politics.[30]

The aesthetics were formalized in Pater's aesthetics as the precise calculations of pleasure received by the critic from the work (the Decadent critic with which this essay began, consuming the culture of the world); Wilde's economies of desire in *Salome* (1893) or *The Picture of Dorian Gray* (1891); Vernon Lee's Psychological Aesthetics; the aesthetico-psychological science of stylistic analysis; and all the literature concerned with distinction, or taste, that distinguished some social groups (women, Celts, Africans, working people) from others.[31] Today, critical theory and cultural studies of 'comparative civilizations' abound, whether the simianization of the Celt or the African, the feminization of the gay male, or the eroticization of the working or 'productive' woman.[32] I have stressed how much these practices of

'Othering' revealed by our contemporary cultural studies were related to taste and the individual 'civilized' man as consumer. Indeed, the literature of the last quarter of the nineteenth century provides a dialectic of individual economic rationality versus its constitutive Other, mass Unreason.

'Comparative civilizational studies' ironically terminated in the first half of the twentieth century, with the fragmentation of Europe into warring states during the two world wars and the consequent defacement of Europe's civilized image. Ideas of civilization and progress were replaced, under North American hegemony, by the language of nation-states and of development. Peter Taylor describes the differences between the progress of civilizations, which was a putative *achievement* and included moral and political progress, and the development of states, which was an economic inevitability:

> First, theories of development are fundamentally state centric in nature: whereas civilizations might 'progress', it is states that 'develop'. Second, the USA becomes the exemplar state. For instance, in the most famous of all development models, states are allocated to different stages but the highest level, the 'stage of high mass consumption', hardly disguises the contemporary United States (Rostow, 1960). Third, unlike progress and civilizations, development and states define a world in which beneficial social change is available to all. The development message was a simple one: follow the required policy prescriptions and you too can be like us – developed, which meant affluent, which meant the United States.
>
> Taylor,'Embedded Statism', p. 5

Clearly the major drift that has taken place between nineteenth-century progress and twentieth-century development is that the ethical and political implications of progress and their accompanying exclusiveness and hierarchy are replaced by the production of goods as commodities for mass consumption. As Disraeli presciently said quite early on, 'the European who talks of progress mistakes comfort for civilization'.[33] While there are still liberals who like to believe that the advantages of 'civilization' in the form of the Good Life follow from economic 'development' as we have known it, the second half of the twentieth century has produced a growing body of critical work on development as a discourse about economic transformation, specifically the integration of 'less developed' states into the global market or, in fully depoliticized terms, the 'war on poverty'.[34] Indeed

Achebe's awesome charge against the *fin-de-siècle* novel *Heart of Darkness* (1899) that Conrad was 'a thoroughgoing racist' in that he supplied no alternative frame of reference for Africa than as setting and backdrop for Europe, could be applied to the development story.[35] 'Can nobody see the preposterous and perverse arrogance in thus reducing Africa to the role of props for the breakup of one petty European mind?', Achebe asked, as an economist might ask whether that great and diverse continent is more than a debit line in the ledgers of the World Bank. Unlike civilization, development is not the culmination of reason in the world (with the flash images from Hegel, Comte and Co.) but economic fiddling with competing *interests* and an increasingly impenetrable discursive preoccupation with the development apparatus itself. This is the transition that Francis Fukuyama has ambivalently heralded as the end of history with its mighty rivalries and spectacular pageantry in economic cost-benefit analyses.[36]

In sum, the man of taste was also the modern, civilized man, and both notions were inextricable from political and economic notions of progress and civilization. He was conceived in dialectical relation not only with women but also with both the internal British barbarian, or the upper-class descendent of feudalism, and the external savage, who did not share his complexity or his self-command. Similarly, the end of taste in aesthetic relativism is also the principle that 'tastes are exogenous' in neoclassical economics, and both are inextricable from economic individualism and twentieth-century ideas of development.

Notes

1 Lawrence Danson, *Wilde's Intentions: the Artist in his Criticism* (Oxford: Clarendon, 1997).
2 Walter Pater, *Selected Writings* (New York: Signet, 1974), pp. 31–52 (pp. 46–7).
3 'Literary History and Literary Modernity', in *Blindness and Insight* (New York: Oxford University Press, 1971), pp. 142–65.
4 Anne McClintock, *Imperial Leather* (New York: Routledge, 1995), pp. 207–31.
5 W. W. Rostow, *Stages of Economic Growth* (Cambridge: Cambridge University Press, 1960).
6 For a fuller account of the shift from production to consumption, see R. Gagnier, 'On the Insatiability of Human Wants: Economic and Aesthetic Man', in *Victorian Studies*, 36 (1993), 125–53.
7 *The Wealth of Nations* (New York: Modern Library, 1937).
8 See *Principles of Political Economy* (London: Penguin, 1988), Books IV & V (1848), and *On Socialism* (Buffalo: Prometheus, 1987).

9 See, for example, *Economic and Philosophical Manuscripts of 1844*, in *The Marx–Engels Reader* 2nd edn, ed. Robert C. Tucker (Norton: New York, 1978), pp. 66–125.
10 *Unto This Last* (1862) (London: Penguin, 1985), pp. 155–228.
11 Stanley Jevons, *The Theory of Political Economy* (London: Macmillan, 1888); Carl Menger, *Principles of Economics* (Glencoe, IL: Free Press, 1950).
12 I have analysed Victorian aesthetics in 'A Critique of Practical Aesthetics', in *Aesthetics and Ideology*, ed. George Levine (New Brunswick: Rutgers University Press, 1994), pp. 264–83; in 'Productive Bodies, Pleasured Bodies: On Victorian Aesthetics', in *Women and British Aestheticism*, eds Kathy Psomiades and Talia Schaffer (Charlottesville: University of Virginia Press, 1998); and in 'Gender Values in Victorian Aesthetics', in *Victorian Sexual Dissidence*, ed. Richard Dellamora (Chicago: University of Chicago Press, 1998).
13 'The Political Economists of Art' in *Unto This Last* (London: Dent, 1932), pp. 1–106.
14 'Art Under Plutocracy', in *Political Writings of William Morris*, ed. A. L. Morton (New York: International Publishers, 1973), pp. 57–85 (pp. 57–8).
15 See *The Painter of Modern Life and Other Essays* (London: Phaidon, 1966), pp. 28–9 and *My Heart Laid Bare and Other Prose Writings* (London: Soho, 1986), pp. 175–210, 213.
16 *Against Nature* (Harmondsworth: Penguin, 1982), p. 27.
17 George Moore, *Mike Fletcher* (New York: Garland, 1977).
18 *The Picture of Dorian Gray*, in *The Portable Oscar Wilde*, ed. Richard Aldington (Harmondsworth: Penguin, 1978), p. 228.
19 See Psomiades and Schaffer, *Women and British Aestheticism*.
20 *The Idea of Progress: a Collection of Readings*, selected by J. Teggart; intro. by George H. Hildebrand (Berkeley: University of California Press, 1949), p. 434.
21 The economic historian David Mitch has discussed how under the influence of Francis Galton's eugenics, and utilizing the principle of non-competing groups, economics similarly moved from essentialist definitions of homogeneous workers to inherently diverse species of workers. Mitch sees this as increasing élitism – 'Victorian Views of the Nature of Work and Its Influence on the Nature of the Worker' (paper presented at the University of California, Santa Cruz, August, 1994).

    For a detailed history of biologism in Victorian theories of progress, see George Stocking, *Victorian Anthropology* (New York: Free Press, 1987) and Peter J. Bowler, *Evolution: The History of an Idea* (Berkeley: University of California Press, 1989), p. 288.
22 *Economic and Philosophical Manuscripts of 1844*, in *The Marx–Engels Reader*, ed. Tucker, pp. 95–6. See also Stocking, *Anthropology*, pp. 234–5, and James Eli Adams, *Dandies and Desert Saints: Styles of Victorian Masculinity* (Ithaca: Cornell University Press, 1995).
23 For distinctions between savagery, barbarism and civilization, see Lewis H. Morgan, *Ancient Society* (London: Henry Holt, 1877); George Stocking, Jr, *After Tylor* (Madison: University of Wisconsin Press, 1995). For an analytic account of the terms, see R. G. Collingwood, *The New Leviathan: or Man,*

        Society, Civilisation, and Barbarism, rev. edn (Oxford: Clarendon Press, 1992). For the most influential of recent theories of civilizing, see Norbert Elias, *The Civilising Process*, 2 vols (Oxford: Oxford University Press, 1978; 1982).
24 P. J. Taylor, 'Embedded Statism and the Social Sciences: Opening Up to New Spaces', *Environment and Planning A*, 28 (1996), 1–12 (pp. 4–5).
25 See especially Stocking, *Tylor*.
26 See Teggart *Progress*, p. 336. For theories of progress, see also Sidney Pollard, *The Idea of Progress* (London: Watts, 1968); Charles Van Doren, *The Idea of Progress* (New York: Praeger, 1967); W. R. Inge, *The Idea of Progress* (Oxford: Romanes Lecture, 1920).
27 In Teggart, *Progress*, pp. 405–6.
28 *Ibid.*, p. 393.
29 For the importance of attention to popular discourse and local usages, see Paul Keating, 'Cultural Values and Entrepreneurial Action: The Case of the Irish Republic', in *Culture in History: Production, Consumption, and Values in Historical Perspective*, eds Joseph Melling and Jonathan Barry (Exeter: University of Exeter Press, 1992), pp. 92–108.
30 See R. Gagnier, 'Is Market Society the *Fin* of History?', in *Cultural Politics at the Fin de Siècle*, ed. Sally Ledger and Scott McCracken (Cambridge: Cambridge University Press, 1995), pp. 290–310.
31 See *ibid.* and R. Gagnier in Levine (ed.), *Aesthetics* (1994).
32 This literature is growing at a rate beyond comprehensibility, but one might begin with McClintock, *Imperial Leather* (1995) and Kelly Hurley, *The Gothic Body* (Cambridge: Cambridge University Press, 1996).
33 Benjamin Disraeli, *Tancred* (London: Peter Davies, 1927), p. 233. Also quoted in W. R. Inge, *The Idea of Progress*, p. 29.
34 See James Ferguson, *The Anti-Politics Machine: 'Development', Depoliticization, and Bureaucratic Power in Lesotho* (Minneapolis: University of Minnesota Press, 1994), pp. 3–21.
35 Chinua Achebe, 'An Image of Africa: Racism in Conrad's *Heart of Darkness*', in *Heart of Darkness* (New York: Norton, 1988), pp. 251–62.
36 *The End of History and the Last Man* (New York: Avon, 1992).

Index

Achebe, Chinua, 236
Achurch, Janet, 211
actresses, 10, 208–21
Adventists, 89
aesthetics, 9–10, 20, 47, 57, 66, 129, 134, 139–40, 147, 157, 176, 190, 208–9, 221–38
agriculture, 91–113, 225
Ainsworth, William Harrison, 9, 127, 142
Aitken, John, 58
almanacs, 8, 91–113
ancien régime, the, 31–2
anthropology, 231, 232, 237
apocalypse, the, 48, 86, 88
Apostles, the, 114–25
Archer, William, 211–12, 221
Aristotle, 67, 70, 152
Arnold, Matthew, 1–2, 5, 7, 13–28, 32–3, 63–79, 180
  Culture and Anarchy, 1–2, 8, 32, 66, 70, 75–7, 184, 190, 192, 226
  Literature and Dogma, 70, 72, 74, 76
  On the Study of Celtic Literature, 71
Aryans, 8, 66, 71, 74–6
astrology, 93, 112
astronomy, 101, 171
Augustine, St, 69
Austen, Jane, 165, 199–201, 205–6, 226
autobiography, 9–10, 15, 19, 162–78, 208–21

Ball, John, 30
Barbauld, Anna, 197–201
Baudelaire, Charles, 11, 143, 228
Beardsley, Aubrey, 223, 228
beauty, 47–8, 58–9, 151–2, 225
Beerbohm, Max, 228
Beeton, Mrs Isabella, 49, 61
Bell, Florence, 215
Benjamin, Walter, 55

Bennett, Arnold, 16
Bentham, Jeremy, 134, 136, 144, 234
Bernhardt, Sarah, 221
Bible, the, 8, 63–79, 80, 101, 163, 167, 174, 181
biography, 20, 130–1, 167
biology, 77, 159
Blackburn, Vernon, 221
Blake, William, 81, 106
Blakesley, J. W., 118, 123, 125
'bloody Sunday', 42–3, 45
body, the, 151, 191, 218, 220, 225, 228
bourgeoisie, the, 31, 36, 131, 183, 190
Brewster, David, 175–6
Bridges, Robert, 33, 58
Brontë, Charlotte, 226, 227
Brontës, the, 226
Brougham, Lord, 164–5
Browne, Montagu, 158
Bulwer (-Lytton), Edward, 9, 88, 127, 128, 130, 142, 227
Bunsen, Baron Christian, 63–4, 66, 70–1, 74
Burnouf, Émile, 71, 74, 75

Cadell, Tom, 2
Caillie, René, 116–17
calendar, the, 91–113
Cambridge University, 6, 8–9, 16, 65, 68, 114–25, 120, 161, 175, 206
capitalism, 4, 132, 182–4, 188
Capper, John, 51
caricature, 98, 104, 106, 108, 112, 128
Carlyle, Thomas, 7, 13–28, 64–6, 75, 77, 80–1, 87–8, 124, 148, 160
Carpenter, William, 59
Carter, Robert Brudenell, 49
Cassell, John, 111, 113
Celts, the, 70, 75
Chambers, Robert, 106, 111
Chambers brothers, 99, 101
chap-books, 98, 106
Chartism, 83, 129

240 Index

Chesterfield, Lord, 46
childhood, 174–5, 198
chivalry, 9, 150–1, 155
Cholmondeley, Mary, 210
Christianity, 8–9, 21, 63–79, 80–113, 119–20, 146–61
civilization, 10–11, 33, 67, 71, 180–1, 224, 226, 229–32, 235–8
Clapperton, Hugh, 116
Cobbe, Frances Power, 165, 168, 173
Cobbett William, 14, 104
Colenso, John, 65
Coleridge, Samuel Taylor, 65, 116
Collins, Wilkie, 52, 140
colonialism, 50, 155, 157, 180
comedy, 98, 132–3
Comte, Auguste, 18, 236
Conder, Eustace R., 53
Condorcet, Marquis de, 232
Conrad, Joseph, 236
consumerism, 3–4, 10–11, 224–6, 228, 235
consumption, 155, 159, 223–6, 228–9, 235
Courvoisier, B. F., 127
Craig, Edy, 219–21
Craig, Gordon, 218, 219, 221
crime fiction, 9, 126–45
Crimean War, 32, 37
criminals, 126–45
Cruikshank, George, 92, 99–100, 104, 106, 108, 111, 113
Crystal Palace, 161
cultural capital, 129, 188–90, 192
cultural criticism, 13–28
cultural history, 8, 10, 22, 131
Cultural Studies, 5–7, 12, 137, 144, 182, 234–5
cultural theory, 4, 6
culture, *see* high culture, mass culture, popular culture, Teutonic culture

dandy, the, 228, 237
Darwin, Charles, 10, 77, 152–3, 179–81, 184, 187, 192, 223, 231
  Origin of Species, 65, 159, 203
day-book, the, 91–113
Decadence, 209, 222–3, 229, 234
democracy, 2, 32–3, 59, 149

desire, 223–4, 228, 234
Dickens, Charles, 9, 52, 80–90, 126–45, 176, 179–94, 226–7
  on acting, 140
  All the Year Round, 61, 84, 143, 180
  'The Amusements of the People', 137
  Bleak House, 48, 86, 127
  Dombey and Son, 84–5, 89
  The Frozen Deep, 140
  Great Expectations, 47
  Hard Times, 127, 137
  Household Words, 51, 144
  Oliver Twist, 9, 126–45
  Our Mutual Friend, 10, 47–8, 51, 179–94
  The Pickwick Papers, 85, 96–7, 112, 130
  Sketches by Boz, 127, 129, 134, 144
  'The Spirit of Fiction', 127
  A Tale of Two Cities, 48, 86, 127
  The Uncommercial Traveller, 84, 86–7, 89
disciplinarity, 13–28
Disraeli, Benjamin, 30, 38, 227, 235
  Sybil, 51
  Tancred, 238
Dixon, Ella Hepworth, 210
Dodsworth, Reverend William, 81
domesticity, 49, 167, 169–70, 172–3, 199, 202, 228
Douglas, Mary, 46, 50, 60
Doyle, Sir Arthur Conan, 53
Drummond, Henry, 82
Duchamp, Marcel, 55
Duse, Eleanora, 211–12, 221
dust, 8, 46–62

earnestness, 87, 189–90, 193, 222
Eastern Question Association, 32, 39
economics, 11, 22, 24, 130, 187, 189, 222–38
Eden, Emily, 50
Edgeworth, Francis Ysidro, 234
Edgeworth, Maria, 176, 200–1
education, 179–94, 190–1, 194, 227
Eliot, George, 18, 77, 201, 206, 226–7
  Middlemarch, 59
  translation of *Life of Jesus*, 65

Index   241

Eliot, T. S., 16, 20
Emerson, Ralph Waldo, 64
empire, 30, 37, 50, 68,78, 83, 117,
    148, 154, 159, 180–1, 231, 234
Engels, Friedrich, 31, 224
English literature (as academic
    subject), 4–6, 13–28, 65, 197
Enlightenment, the, 11, 81, 85, 143,
    232
    Scottish Enlightenment, 174
ethnography, 8, 70, 76–7
eugenics, 237
Evans, Jean, 204, 207
evolution, 75, 77, 159, 180–1, 187,
    192, 223, 232, 234
Ewing, J. H., 159

fairy tales, 46
femininity, 10, 75, 163, 166, 169–70,
    173–4, 212
feminism, 163, 165, 168, 173
Feuerbach, Ludwig, 66, 70
    Essence of Christianity, 65
fin de siècle, 222–3, 228–9, 234, 236
Fox, W. J., 30, 38, 45, 136, 137
Francophobia, 31, 44
French Revolution, 81, 86, 92, 129
Freud, Sigmund, 51

Galton, Francis, 237
Gardiner, Grace, 50
Gaskell, Elizabeth, 50, 78, 226–7
genre, 13–28, 91–113, 126–45,
    162–78, 196, 208–21
ghost stories, 195–207
Gibbon, Edward, 66, 181
Gissing, George, 41, 42
Gladstone, William, 30
Glasgow University, 65
Gobineau, Count Joseph Arthur de,
    66, 71–4
Godwin, William, 91
Goethe, Johann Wolfgang von, 64,
    72, 230
Golden Jubilee, 8, 30, 39–42
Gosse, Philip, 59, 147, 153, 158, 161
Grand, Sarah, 210
Greig, Samuel, 170
Grundy, Sydney, 210

Haggard, H. Rider, 234
Hallam, Arthur, 54, 116–17, 124
Harkness, Margaret, (John Law), 41–4
Harrison, Jane, 202–3, 206
Hartley, John, 111
Haweis, Mrs (M. E.), 228
Heath, Henry, 113
Hegel, Georg Wilhelm Friedrich , 64,
    234, 236
hegemony, 9, 24, 27, 159, 182, 235
Henley, W. E., 48, 61
heroism, 52, 66, 117, 127, 135, 138,
    150, 164, 183, 186, 228
Hetherington, Henry, 94
high culture, 2, 5, 12, 136, 144, 182
Hogarth, William, 96
Hone, William, 94–5, 99, 104, 106
Hood, Robin, 32, 132
Hood, Thomas, 64
Hopkins, Gerard Manley, 33, 54, 57–8
Howitt, Mary and William, 99, 106,
    113
Hunt, Leigh, 119
Huxley, Thomas Henry, 65
Huysmans, J. -K., 228

Ibsen, Henrik, 210–11, 215–16
illustrations, 96, 106, 108, 112, 218
imagination, 47–8, 55–6, 60, 72, 117
imperialism, 9, 68, 155
individualism, 23, 168, 224, 229, 236
Indo-Europeans, 66, 70, 72, 74–6
industrialization, 2, 4, 22, 25, 50, 55,
    91, 94, 163, 180
Irving, Sir Edward, 80, 82, 87
Irving, Henry, 216–18

James, Henry, 215
Jeffrey, Francis, 164
Jewishness, 63–79, 82, 132, 144
Johns, Reverend Charles Alexander,
    158–9
Jourdain, Eleanor, 203–4

Kant, Immanuel, 56, 60, 64, 143, 234
Kean, Edmund, 12, 218
Keats, John, 120
Keble, John, 93, 152
Keith, Reverend Alexander, 82–3

Kemble, John, 124–5
Kingsley, Charles, 8, 63–79, 146–61
Kipling, Rudyard, 202
Knight, Charles, 94, 99

labour, 10, 31, 98, 125, 148, 194, 224–5, 228, 230–1
Laing, Major, 116, 124
Land and Labour League, 30–1, 36
Laplace, Pierre Simon, 164–5
Lardner, Dionysius, 94, 100, 101, 112
Leavis, F. R., 5–6, 22, 136, 144
Lee, Vernon, 234
Lessing, Gothold Ephraim, 66, 70
Lewes, G. H., 59–60, 62, 147
liberalism, 24–5, 235
literacy, 2, 112, 129
literary criticism, 13–28, 128, 136–7, 147
literary history, 199, 236
literary studies, 5–7, 12–28, 137, 144, 182
literary theory, 5–6, 27, 77, 144, 234
Lloyd, Marie, 39
London, 41–3, 48, 50, 64, 171, 174–5, 215
London University, 120, 124, 172
Lowth, Robert, 64
Lubbock, John, 152
Lyell, Charles, 180

MacKintosh, Mabel, 53
Macaulay, Thomas Babington, 21
Macready, William, 214
McGonagall, William, 40
McPherson, J. G., 47–8, 61
Maginn, William, 127
Malet, H. P., 49
Malthus, Thomas Robert, 22, 225
Marcet, Jane, 164
Marie Antoinette, Queen, 204
Martin, Helena Faucit, 214
Martin, John, 81, 88
Martineau, Harriet, 168, 177
Marx, Karl, 31, 36, 182–3, 185, 188, 190, 194, 224, 231
 Capital, 182, 185
 The Eighteenth Brumaire, 10, 183–4, 194

Marxism, 2, 77, 139, 182
masculinity, 148, 151–?, 161, 165, 196
mass culture, 4, 9, 106, 130, 136, 142
mass media, 12, 130
materialism, 2, 189–90, 202
mathematics, 9, 162–78
Maurice, F. D., 116, 119, 121–2, 125
Maxse, F. A., 33
Mayhew, Henry, 48, 51
Mazzini, Joseph, 34
melodrama, 3, 9, 130–1, 136–9, 145
Meredith, George, 33–4, 44
Meynell, Alice, 195
microscope, the, 48–9, 59–60, 156
millenarianism, 8, 80–90
millennium, the, 8, 80–90
Mill, John Stuart, 13–28, 180, 224–5
Miller, Hugh, 161
Miller, William, 83
Milnes, Monckton, 118, 124–5
Milton, John, 34–35, 167
Minerva Press, 2
Moberley, Charlotte, 203–4
modernity, 4, 11, 88, 129, 143, 222–38
Modern Languages Association, 13, 27
monarchy, the, 7, 29–45
 French monarchy, 82
Moore, George, 228, 237
moralists, 13–28
Morgan, Lady, 173
Morris, William, 37–9, 227
Morrison, Arthur, 42
Most, Johann, 39
music hall, 39
Müller, Max, 10, 179–81, 185, 188, 192

Napoleonic Wars, 65
national identity, 31, 174–6
natural history, 9, 98, 112, 146–61, 171
naturalism, 42, 72
naturalist, the, 9, 59, 151, 153
nature, 9, 24, 47, 57–8, 60, 72, 147, 146–61, 163, 213, 225
New Criticism, 6, 7, 16–17, 20–1

Newgate Calendar, 126, 131, 133–6
Newgate debate, 9, 126–45
Newgate fiction, 9, 126–45
Newgate prison, 126, 133, 136
New Woman, the, 208–21
Newman, John Henry, 18–19, 63, 76
Nietzsche, Friedrich Wilhelm, 24
Nightingale, Florence, 49
non-fiction prose, 13–28, 130

Oldfield, Ann, 218
Oliphant, Margaret, 165, 173, 175–6, 199, 201–2
Opie, Amelia, 198–9, 201
originality, 9, 163–4, 176, 213
Ouida, 53–4
Oxford University, 5, 9, 65, 96–7, 119

Paine, Tom, 99
Palgrave, Francis Turner, 40
Palmer, John, 3
Paris Commune, the, 7, 31–4
Parker, John, 94
parody, 141, 145, 183, 192
Pasteur, Louis, 48
Pater, Walter, 18, 222–3, 234
patriotism, 44, 104, 116
Pène, Henri de, 55
periodicals, 25–6, 108, 129–30, 146, 159
philosophy, 8, 15, 18, 20, 147, 154
physiology, 159, 230, 234
Pinero, Arthur Wing, 210
Plato, 67, 70, 77
pleasure, 9, 135–7, 139–40, 142, 144–5, 151, 154, 157, 228–9, 234
political economy, 24, 224–8, 230–1
Pollock, Juliet, 208
popular culture, 2, 4–5, 8–9, 12, 39, 59, 80, 91–113, 126–45
postmodernism, 128, 137
post-modernity, 88
post-structuralism, 21, 135, 137
professionalism, 25, 152, 168, 194
progress, 55, 70, 98, 222–38
psychoanalysis, 13, 136
Pusey, Edward, 63, 65

Quiller-Couch, Sir Arthur, 16, 27

radicalism, 29–45, 91–113, 124, 181, 211
reading public, the, 15, 93
Reade, Charles, 218–19
realism, 127, 131, 145, 187, 190, 192–3
Reed, Charles, 53
Reform Bills, the, 2, 32, 124, 129
Renaissance, the, 29, 93
Renan, Ernest, 64, 71–3, 74–5, 78
Republicanism, 8, 29–45, 183
Ricardo, David, 224
Ritchie, Anne Thackeray, 10, 195–207
Robertson, Graham, 216
Robins, Elizabeth, 210, 211, 214, 215, 216
Romance, 128, 136, 147, 196
Romanticism, 13, 28, 45, 72–3, 123
Rossetti, Christina, 89
Rossetti, Maria, 81
Ruskin, 7, 13–28, 52, 56–7, 104, 149, 152, 225, 227

St John, Christopher, 219–21
Saintsbury, George, 15
Sand, George, 205
satire, 98, 109, 141, 210
Saussure, Ferdinand de, 6
Schiller, F. C., 64, 70, 78
science, 8–9, 46–8, 56, 58–60, 65–6, 69–70, 75–7, 146–78, 179
Scott, Sir Walter, 2, 176
Scottishness, 174–6
Second Coming, the, 80–90
Semites, 8, 63–79
sensibility, 21, 23
serialization, 129–30
sexuality, 143–4, 223
Shakespeare, William, 26, 72, 85, 214, 219
  Romeo and Juliet, 208, 214
Shaw, George Bernard, 211, 221
Shelley, Mary, 226
Shelley, Percy Bysshe, 34, 35, 39, 116–18, 121
Sheppard, Jack, 126–7, 131
Smith, Adam, 224
Smith, Joseph, 84, 86
Smollett, Tobias, 106

social sciences, 15, 231–2
socialism, 29, 25, 236
sociology, 8, 232, 234
Society for the Diffusion of Useful
    Knowledge (SDUK), 94–5, 99,
    101, 164–5
Society for the Propagation of
    Christian Knowledge (SPCK),
    94–5, 99, 101–2
Socrates, 67, 70, 77
Solomons, Ikey, 132
Somerville, Martha, 162–78
Somerville, Mary, 9–10, 162–78
Southcott, Joanna, 81, 83, 85
Speke, Captain J. H., 181
Spencer, Herbert, 18, 229–32
sportsmanship, 151, 157, 161
Stationers' Company, 94–5
Steele, Flora Annie, 50
Sterling, John, 116–17, 121–2, 125
Stevenson, Robert Louis, 196–7
Stoker, Bram, 55, 218
Strauss, David Friedrich, 65, 66
subjectivity, 8, 139, 148, 152, 159, 225
Surr, Elizabeth, 146
Swinburne, A. C., 34–7, 45
symbolism, 21
Symons, Arthur, 212–13

taste, 111, 130, 144, 188–9, 191, 226,
    229, 231–2, 234–6
telescope, 60
temperance, 96
Tennent, Sir James Emerson, 140
Terry, Ellen, 216–21
Tennyson, Lord Alfred, 8–9, 31, 44,
    54, 58, 114–25, 180
Tennyson, Charles, 122
Teutonic culture, 63–4, 66, 69–70, 77
Thackeray, William Makepeace, 126,
    128–9, 142, 226
theology, 8, 63–79, 148, 153
Tilt, Charles, 113
Tolstoy, Leo, 210
Toryism, 17, 38, 104, 106, 127
Trades Union Congress, the, 31
Tree, Beerbohm, 216
Trench, Richard Chevenix, 122, 123, 124

Trollope, Anthony, 226–7
Tupper, Martin, 44
Turner, J. M. W., 113
Turpin, Dick, 126–7, 131
Tyndall, John, 48, 55–7, 60

Utiltarianism, 22, 116

vampire, the, 55, 222, 223
Vestris, Madame, 208
Victoria, Queen, 1, 7, 10, 29–31, 36,
    38, 40–1, 129, 164
Victorianism, 29–45, 170
Victorian poetry, 3, 5, 9, 11–12,
    114–25, 130, 144
Victorian theatre, 3–4, 10, 130–1,
    138–9, 143, 208–21
vision, 46–62

Waddington, Frances, 63
Wallace, Alfred Russel, 46, 57–8, 60
Ward, Mrs Humphry (Mary), 59, 69–70
Ward, John, 83, 85
waste, 47, 51
Wellington, Duke of, 120
Wesley, Charles, 81
Whewell, William, 166–7, 177
Wilberforce, Samuel, 65
Wilde, Oscar, 18, 222–3, 228, 234
Williams, Raymond, 1, 2, 6–7
    Culture and Society, 1–2, 6, 7, 13–28
    The Long Revolution, 194
    Politics and Letters, 23
    'structures of feeling', 10, 139, 181–2
Wollaston, William, 175
women's writing, 9–10, 162–78,
    195–221
Woolf, Virginia, 10, 195–6, 199,
    205–7, 220
Wordsworth, Christopher, 120–1
Wordsworth, William, 73, 116, 118,
    121, 125
Wroe, John, 83

year-books, 8, 91–113
Yonge, Charlotte, 150, 204

Zimmern, Helen, 211–12
Zola, Émile, 210

KING ALFRED'S COLLEGE
LIBRARY